E. M. Lloyd

Vauban, Montalembert, Carnot

Engineer Studies

E. M. Lloyd

Vauban, Montalembert, Carnot
Engineer Studies

ISBN/EAN: 9783337060794

Printed in Europe, USA, Canada, Australia, Japan

Cover: Foto ©berggeist007 / pixelio.de

More available books at **www.hansebooks.com**

VAUBAN, MONTALEMBERT, CARNOT:

ENGINEER STUDIES.

BY

E. M. LLOYD,

MAJOR, ROYAL ENGINEERS; LATE PROFESSOR OF FORTIFICATION AT THE
ROYAL MILITARY ACADEMY, WOOLWICH.

WITH PORTRAITS.

LONDON—CHAPMAN AND HALL,

LIMITED.

1887.

(All Rights reserved.)

PREFACE.

THE three essays which form the chief part of this volume, have already appeared in the *Quarterly Review*, the *Royal Engineers' Journal*, and the *United Service Magazine*, respectively. Some portions of the Introduction were included in the new *Text-book of Fortification* for the use of the Cadets of the Royal Military Academy, issued in 1878; and the last paper, on modern forts, was published in the *Journal of the Royal United Service Institution*, in 1882.

They are now brought together as illustrating successive stages in the history of fortification and fortress warfare. It is hoped that they may furnish some useful clues to those who care to study that subject in the only way in which it can be studied to much purpose, by going to the fountain heads. At all events, the great men of whom some account is here given, deserve to be something more than names, even to those who are not military students.

The author takes this opportunity of thanking the proprietors of the several magazines above mentioned, for kindly assenting to this republication.

CONTENTS.

CHAPTER I.
	PAGE
INTRODUCTION	1

CHAPTER II.
VAUBAN	60

CHAPTER III.
MONTALEMBERT	98

CHAPTER IV.
CARNOT	155

CHAPTER V.
MODERN FORTS	200
INDEX	237

ENGINEER STUDIES.

CHAPTER I.

INTRODUCTION.

ANY one who compares an old castle with a modern fortress, who turns, say, from the eastern to the western heights of Dover, must be struck not only by the many points of difference between them, but also by the singular course of development which these differences indicate.

The new works are neither so imposing nor so elaborate as the old, and instead of advancing steadily upon one line, the art of defence seems to have changed direction, and in some respects to have gone backward. The reason is that fortification is peculiarly dependent upon other arts. It is the making of armour to protect men from the weapons of others while allowing them to use their own, and must adapt itself therefore to all the changes of weapons, instead of taking its own course.

Corresponding to the broad division of missiles and hand-weapons, the main elements of all fortification are cover and obstacle. Down to the fifteenth century, handweapons played the principal part in war; missiles were an accessory, largely neutralized by body-armour; and the great aim of military architects was to increase and multiply the passive obstacles which hindered besiegers from coming to close quarters with the besieged. But the enormous development of missile force which has followed on the adoption of gunpowder has thrown hand-weapons into the background. Fortification has had to take a new departure, and the skilful adaptation of cover, so as to combine the utmost effectiveness of the fire of the defence with the utmost immunity from the fire of the attack, has become more and more definitely the problem of the engineer.

The character of mediæval fortification, and the methods of attack and defence, have been described with admirable vividness by M. Viollet le Duc in his "Military Architecture of the Middle Ages."

The walls consisted of curtains and towers; the curtains high enough to make escalade difficult, and with a width of only two or three yards at the top, made up of the battlemented parapet wall with a narrow path behind it; the towers higher and more massive, projecting in front of the curtain, and flanking both the face and the top of it.

"The ancients," says Vegetius, "did not think it well that the wall enclosing a place should be carried straight; and so lend itself to the blows of the battering-rams; but preferred to make it alternately salient and re-entering, and built numerous towers at the angles. Hence if any one attempts to bring up ladders or engines against a wall so constructed, he is hemmed in and assailed not only in front, but also in the flanks and almost in rear."[1]

[1] "De re militari." Bk. iv. ch. 2.

Besiegers consequently preferred to attack acute salients, which, owing to the short range of the engines for hurling projectiles, were necessarily ill-flanked. While such salient angles were avoided as much as possible in fortifying a place, where they were unavoidable the corner tower was made larger and higher than others; its point was made to jut outwards, so as to form a horn or beak, offering greater resistance to the ram or sap, and admitting of better defence from the curtains; obstacles were multiplied in front of it, and arrangements made for retrenching in rear of it in case of its capture.

But while the value of an enveloping fire, and the importance of seeing the walls to their bases was fully appreciated, the facilities which the slight stone parapet gave for a vertical defence made constructors comparatively careless about a strict lateral defence. There was not much difficulty in keeping a watch from the battlements upon what was going on down below, or in pushing over stones and timber; and overhanging wooden hoardings or stone machicolations made this surveillance and defence easier and more thorough. It was only when improvements in the besieger's missile engines made overhanging works less secure from destruction, that it began to be thought worth while to occupy and loophole the lower parts of the towers.

A siege consisted of a series of attempts to get over the walls, through them, or under them. With the first object escalade might be tried, or men might be landed on the top of the walls from high wooden towers (*beffrois*), for which causeways had been made across the ditch. Breaches were usually made by pioneers working under shields (*pavises*), and supported by archers, though the battering-ram was occasionally used. Engines (*trébuchets*), slung stones and fire-barrels, which destroyed the hoardings and houses, but their force was only equal to that of a few ounces of gunpowder and they could do nothing against walls. If the soil admitted of it, mine galleries were carried under the walls, either to bring them down,[2] or to open an unsuspected passage into the place.

But all these means were very inadequate to deal with fortresses strong by nature and art, and defended by resolute men. In such cases besiegers were usually driven either to raise the siege, or to rely upon starving out the

---

[2] "If undermining seemeth him good, good workmen that can skill shall be set to work for to delve up the earth, and they shall begin so far that they of within shall not by no way see the men that bear out the earth. And so deep shall the mine be made that it shall pass under the ditches, which shall be undershored with good timber, till that they come to the foundments of the walls or lower; and by this manner of way they shall find the means to enter in, if gainsaid be not done to it. And while this undermining is adoing the wise captain ought not to be still, lest they of within feel or understand by their scoutwatch the said underminers, but shall trouble and vex them with divers and continual assaults, so that the noise, the doing and the bruit shall stop their hearing. For strokes of crossbows thicker than flies, bombards and guns, with the horrible sound of their stones cast against the walls, the noise of the assaulters, the sounds of trumpets, and the fears that they have of them that climb up the ladders shall give them enough to do, and so they shall not be little occupied. Item, and if it hap that the said miners may pierce the walls without they be perceived, they shall soon set on fire all the timber and shores that they have set under the walls, which shall then break and fall down all at once, and thus shall enter the town the men of arms."—Christine de Pisan, "The Book of fayttes of Arms and of Chyvalrye." Translated from the French by W. Caxton. Westminster, 1489. The work was written in the early part of the fifteenth century.

garrison by a blockade, a thing which the conditions of military service in those days made it difficult to maintain for many months.

It was the feudal castles rather than the cities that most often defied capture; cities were more amenable to blockade, and besides their sites had not usually been chosen with a single eye to defence. The castles were placed on commanding bluffs or spurs, with steep slopes, altogether inaccessible by the engines of the besiegers, and were usually approachable only on one side, perhaps by a narrow ridge. The difficulty of breaching or escalading their walls led besiegers to address themselves to the gates, and to endeavour to penetrate these either by force or surprise, in spite of the obstacles accumulated at these points. Hence, when men wished to convey a high idea of the strength of a place, they said it had only one or two gates. "We are struck," says M. Viollet le Duc, "when we study the system of defence adopted from the twelfth to the sixteenth century, with the care taken to guard against surprise; all kinds of precautions are taken to arrest the progress of the enemy, and to embarrass him at every step, by complicated arrangements in the place, and by turns and checks, which it was impossible he could foresee. It is evident that a siege, before the invention of cannon, was never really serious, either for the besieged or the assailants, except when it became a hand-to-hand contest."

Step by step defence and vertical defence are the two leading characteristics of mediæval fortification, and its progress is chiefly to be marked in the increasing skill and intricacy of the arrangements for bringing the enemy to a check again and again, whether he penetrates by the gates or the walls, and in the increasing solidity of construction of the machicolations, and the greater height to which they are raised, so as to be out of range of the engines of the attack. At the keep of Coucy provision was made for two-storied hoarding at a height of nearly 200 feet above the bottom of the moat.

Cannon were made use of at all events before the middle of the fourteenth century, and towards the end of that century bombards of large calibre were introduced, which threw stone shot of several hundred pounds' weight. But though these shot produced great effect on the walls wherever they happened to strike, the badness of the powder and the weakness of the guns made the shooting too uncertain for systematic breaching. Hence, until the middle of the fifteenth century, cannon were of more advantage to the defence than to the attack. The besieged profited by the lighter and handier pieces in repelling assaults, and had little need of the large ones.[3] The besieger had to bring with him a train of pieces, often difficult to obtain and difficult to transport; whereas the old engines could be made on the spot. He usually directed his fire upon the gates rather than upon the walls, and to protect them, it soon became the custom to throw up bulwarks, or boulevards (German *boll-werk*, timberwork), of earth and timber in front of them. This, and

---

[3] Christine de Pisan, upon the authority of "wise knights," estimates the requirements of a castle or town with a garrison of 600 men as follows :—twelve guns, casting stones, two of them great, to break engines, mantlets, &c.; 1000 lbs. of gunpowder, and 600 lbs. of lead to make pellets; 200 gunstones ready made, with plenty of other stone to make more if need be; also trébuchets, bows, crossbows, quarrels, arrows, and hand-weapons.

the slight enlargement of the arrow-slits required to convert them into gun-ports, were the only modifications of construction that artillery for some time gave rise to.

The relative strength of the defence being unimpaired, blockade remained the chief hope of besiegers in dealing with large and well-garrisoned places. Henry V. waited passively for six months before Rouen, as Edward III. had waited before Calais and Rheims. Henry's investment lines would almost compare with Cæsar's lines round Alesia. He had "large trenches excavated between his tents and the walls, a crossbow-shot from the latter, which soon enveloped the town with a continuous contravallation. The earth thrown to the inner side of the ditch formed a parapet, which was made to bristle with spikes. In front of this vallum, to stop the enemy's horse, several rows of pointed stakes were planted. Between the posts, deeply sunken covered ways gave secure communication from corps to corps. Places of arms at intervals, and barracks made with logs and young trees interlaced and covered with sods, formed fresh towns as it were round the town."[4] He threw a bridge over the Seine, about three miles above the town, and as winter drew on he made a second line of circumvallation round his camps, to guard against any attempt at relief. It was like the first line in its general character, flanked at intervals by towers, and garnished with cannon and balistæ.

Henry's army was about 30,000 strong, and was therefore able to maintain a close investment, and to repel the efforts of the garrison to break through. But with

[4] Puiseux, "Siége et prise de Rouen."

smaller forces only a partial blockade was possible. Small detached works, termed indifferently bastilles and bulwarks, and made of timber or of earth, were placed to bar the main roads, instead of forming continuous lines. This was the case at Orleans. The English first took the Tourelles at the south end of the bridge over the Loire, and then they gradually completed a chain of bastilles on all sides of the town. But their numbers, originally only 10,000 and at length reduced to 3,000, were quite insufficient to prevent ingress or egress, or even to guard their own works. As soon as the coming of the Maid had revived the hearts of the garrison, the bastilles on the east and south sides were taken one after another; the Tourelles and the bulwark in front of it were recovered after hard fighting, and the English had to raise the siege.[5]

Timber and hurdlework played the principal part in protecting besiegers from the missiles of the besieged, so long as those missiles were stones and arrows, for at a distance from the walls cover was not needed at all, and near the walls it was chiefly needed against missiles coming from above.[6] But earth began to take their place as cannon came to aid the defence and gunners grew expert. Many of the English leaders in the French wars,

---

[5] Jollois, "Histoire du Siége d'Orléans."

[6] An estimate of the brushwood and timber required for bastilles and covered ways in besieging a strong place is given by Christine de Pisan. It comprises 640 paleboards or panels, fourteen feet long and twelve feet broad, with two trestles to each panel, "garnished with hurdles for to make the alleys and ways to go over." Also about the same number of smaller panels, ten feet high, and twelve feet broad, to make bastilles and bulwarks, each of which should have four gates fortified with towers.

for instance, the Earl of Salisbury, the Earl of Arundel, and the two Talbots, were killed by cannon-shot. At Orleans, as at Rouen, we find mention of trenches made to serve as covered ways from camp to camp, and soon after we see them used as approaches to the edge of the ditch. Earth began also to be employed to cover the besieger's guns, first in casks, and then in large wicker-work gabions, eight or nine feet high, and four or five feet in diameter.

By the middle of the fifteenth century, artillery had made a fresh advance, this time in favour of the attack. The calamities of France and of Charles VII. had drawn the people to the king and the king to the people. His very disasters had set him free from the great nobles, and he chose able men of mean extraction to manage his affairs. They brought his finances into order; they furnished him with troops, both horse and foot, in his own pay—the first beginning of a national army—with which to assert his authority over his vassals; and they gave him, what was no less essential, an artillery against which feudal castles could not hope to hold out. This was chiefly the work of the brothers Bureau. Their most marked improvement was the substitution of cast-iron for stone shot, which allowed the calibre and consequently the size and weight of the guns to be much reduced, while they gained greatly in penetration. But besides improving the quality of the king's cannon, they largely increased their number. "Never king had such a train so well supplied with ammunition and every implement for battering towns, nor so numerous a body of men and horses to draw them," says the continuator of Monstrelet.

When he took the field once more against the English, in 1449, after the five years' truce of Tours, his new strength showed itself in his rapid success. The reconquest of Normandy, involving sixty sieges, was accomplished in a year. The capture of Harfleur, the key of Normandy, had cost the English a month in 1415, and four months in 1440. Charles recovered it in seventeen days, in the depth of a severe winter. "Sixteen large bombards were planted against the walls ... deep trenches of communication were formed .. and covered trenches were carried to the very walls of Harfleur; and these mines and trenches were formed under the direction of Master John Bureau, treasurer to the king, in conjunction with his brother Jasper, grand master of the royal artillery, both very expert and able in the sciences."[7] "It was wonderful," says the same writer elsewhere, "to see their diligence in planning and forming the trenches and mines that were opened at almost every siege during this expedition; for to say the truth, there was scarcely any place that surrendered on capitulation but what might have been won by storm."

The moral effect of such success was soon enough by itself. The castle of Montereau surrendered as soon as the heavy cannon were pointed against its walls. Bayonne held out for months against ordinary guns, "but when they knew that the bombards were coming, they of the town asked for terms."[8]

Louis XI. took care to retain the advantage, which his

[7] Monstrelet (continuator), iii. 27.
[8] Napoleon III., "Le passé et l'avenir de l'Artillerie," ii. p. 100.

INTRODUCTION. 11

father had secured, of a good and numerous artillery; and the brothers Bureau found successors to carry on their work. Guns of cast bronze had long been gradually superseding the built-up guns. They now began to have trunnions cast upon them, strong enough not only to carry their weight, but to bear the shock of discharge; so that the guns could travel on their firing carriages, and be brought into action at once, and when in action they could be served and pointed much more rapidly than of old.[9]

The effect of these improvements was strikingly shown when Charles VIII. invaded Italy, in 1494, and men first began then generally to realize how urgent and how considerable were the changes called for in fortification to meet the new means of attack.

In Italy, while learning, art, and science were being pursued with all the ardour of the opening Renaissance, war and its concerns were left mainly to the Condottieri, whose chief study it was to live and let live, to earn their pay without injury to one another. The contrast between the Italian and the French siege artillery is described by Guicciardini. The former "were of so large a size that, on account of the little experience of the artillerymen, and clumsiness of their carriages, they were moved from place to place very slowly and with great difficulty, and for the same reason were very unhandy when placed against the walls of a town. The intervals between the firings were so long that a great deal of time was lost, and little progress was made in comparison to what we see in our days. This gave time to the besieged to cast up ramparts and fortifi-

---

[9] Napoleon III., op. cit., p. 113.

cations behind the breaches at their leisure. But notwithstanding all these impediments, the violence of the saltpetre, of which gunpowder is made, was such that when these instruments were set on fire the balls flew with so horrible a noise and stupendous force, even before they were brought to their present perfection, that they rendered ridiculous all the instruments so much renowned, invented by Archimedes and others, and used by the ancients in sieges of towns. But now the French brought a much handier engine, made of brass, called *cannon*, which they charged with heavy iron balls, smaller without comparison than those of stone made use of heretofore, and drove them on carriages with horses, not with oxen, as was the custom in Italy; and they were attended with such clever men, and on such instruments appointed for that purpose, that they almost ever kept pace with the army. They were planted against the walls of a town with such speed, the space between the shots was so little, and the balls flew so quick and were impelled with such force, that as much execution was done in a few hours as formerly in Italy in the like number of days."[1]

They were not often called on to put their powers to the proof, for Charles met with little resistance on his road to Naples. No army stood in his way, and places opened their gates at the first summons. Such exceptions as there were proved warnings to others. Monte San Giovanni, on the borders of the kingdom of Naples, was "a place by situation strong, well fortified, and provided with a numerous garrison: for there were 300 foreign foot, and 500 of the inhabitants determined to defend

---

[1] "History of the Wars in Italy" (Goddard's translation).

themselves to the last, which made people imagine the French would be detained here for some days. But after firing the cannon for a few hours, they gave the assault in the king's presence . . . . with so much bravery that they overcame all difficulties and took it by storm the same day, and prompted by their own natural fury, and also to set an example to others not to make any opposition, made a vast slaughter, and after perpetrating all sorts of barbarities, they exercised their cruelties against the edifices by setting them on fire. This manner of making war not having been practised in Italy for many ages filled the whole kingdom with vast consternation."[2]

In the long struggle between foreigners for predominance in Italy which followed upon Charles VIII.'s invasion, the Italians had large experience of the "new and bloody ways of making wars" of which Guicciardini complains. They had hardly had time to learn the lesson taught by the new artillery, when another potent instrument of attack had to be reckoned with. In 1503, when Gonsalvo of Cordova besieged the castles of Naples, his engineer, Peter of Navarre, mined the walls, and blew them down by gunpowder, and so took the castles, "to his own great reputation, and to the great astonishment of all men." He is said to have seen such mines employed by the Genoese, in 1487, at the siege of Sarranello, and the idea may be traced further back.[3]

[2] Guicciardini, op. cit.

[3] Colonel Augoyat quotes from a work by Taccola of Sienna, "De Machinis," written in 1449: "Fiant caverne per fossores penetrautes usque sub medium arcis. Ubi hauserint strepitum pedum sub terrâ, ibi faciant cavernam latam admodum furni, in eam immittuntur tres aut quatuor vigites sursum apertos plenos pulvere bombarde; inde ab ipsis vigetibus ad portam caverne ducitur

It seems probable [4] that it developed itself gradually, powder having been first employed merely to knock down the wooden props of the old mine galleries, in place of burning them. It was a matter of some difficulty to get these timbers to burn in the atmosphere of long underground passages, and especially to get them to fall at the same moment. But the effect of powder would not be considerable until it was skilfully applied, in chambers not too large, well tamped, and suitably placed; and Peter of Navarre was the first man so to apply it.

How much contemporaries were impressed with this new danger may be seen in Machiavelli's writings. Speaking of places strong by nature, he says that for this they must in these times either be like Mantua, surrounded with fens, or like Monaco, perched on a rock; "for those that stand upon hills that be not much difficult to go up, be nowadays, considering the artillery and the caves (scil. mines), most weak." The remedy for the latter is to build in the plain, "and to make the ditch that compasseth the city so deep that the enemy may not dig lower than the same where he shall not find water, which only is enemy to the caves." [5]

This seems to strike the keynote of modern fortification. The bold heights whose steep slopes gave security against the cat, the *beffroi*, and the *trébuchet*, can give no such security against the cannon-ball; and the new terrors of the mine make it the more urgent to seek safety

funiculus sulphuratus. Qui, obturatâ portâ caverne lapidibus et arenâ et calce, accendatur. Sic ignis pervenit ad vigites, et concitatâ flammâ, ara in medio posito comburitur." (*Spectateur Militaire*, 1846.)

[4] " Le passé et l'avenir de l'Artillerie," ii. 129.
[5] " Dialogues on the Art of War " (Withorne's translation, 1573).

not by rising above the ground, but by sinking into it.

But besides the question of site, there was the question of construction, which was urgent enough to engage the attention of statesmen as well as soldiers, and to occupy the minds of some of the best artists of the Renaissance. Indeed, the work of fortress building was far from being a distinct profession at that time. It fell to civil architects, as one branch of their business, and was so undertaken by Bramante, Michael Angelo, San Michele, and others. As the term engineer implies, it was for the machines and contrivances required in sieges, rather than for the designing of permanent works, that the need of a special class of men was first felt. Leonardo da Vinci, in his letter commending himself to L. Sforza for employment, lays chief stress on his cunning contrivances.

"If the walls be made high," says Machiavelli, "they be too much subject to the blows of the artillery; if they be made low they be most easy to scale." How to escape this dilemma was the problem to be solved, and the solution was sought in three different directions: first, by retrenching the wall; secondly, by strengthening it; thirdly, by shielding it by works outside of it.

Machiavelli, who had gone practically into the question, and had superintended the construction of the new citadel at Pisa, in 1511, by Giuliano and Antonio da San Gallo, recommends the first of these courses. "If thou makest the ditches on the outside thereof for to give difficulty to the ladders, if it happen that the enemy fill them up (which a great army may easily do), the wall remaineth taken of the enemy. Therefore . . . the wall ought to be made high and the ditch within . . . at least twenty-two and a half yards broad and nine deep; and all the earth that is digged out for to make the ditch must be thrown towards the city and kept up of a wall that must be raised from the bottom of the ditch and go so high over the town that a man may be covered behind the same, the which thing shall make the depth of the ditch the greater. In the bottom of the ditch within every 150 yards there would be a slaughter-house (*casa mata*) which with the ordnance may hurt whomsoever should go down into the same; the great artillery that defend the city are planted behind the wall that shutteth the ditch, because for to defend the outer wall, being high, there cannot be occupied commodiously other than small or mean pieces. If the enemy cometh to scale, the height of the first wall most easily defendeth thee; if he come with ordnance it is convenient for him to batter the outer wall; but it being battered, for that the nature of the battery is to make the wall to fall towards the part battered, the ruin of the wall cometh, finding no ditch that receiveth and hideth it, to redouble the profundity of the same ditch, after such sort that to pass any further it is not possible, finding a ruin that withholdeth thee, a ditch that letteth thee, and the enemy's ordnance that from the wall of the ditch most safely killeth thee."

The Florentines had had ample experience of the force of this method in their repeated attempts to recover Pisa, which were foiled again and again by the Pisan retrenchments, and after fifteen years only succeeded by famine in 1509. But it was exhibited even more strikingly in the defence of Padua in that year, when the Pope, the Emperor and the kings of France and Spain were all leagued together against Venice. The Venetians had made a continuous ditch and rampart inside the walls, flanked by small casemates, just as Machiavelli recommends. They had also backed the walls with well-rammed earth, and had made bulwarks in front of them at various points, armed with artillery to flank the outer ditch. They had mine chambers ready to blow up any of the works that might be taken by the enemy. The accumulated defences were so formidable that, after making a breach and trying an assault, Maximilian raised the siege.

But there was seldom room for an inner ditch and rampart all round a place. In most cases it could only be employed for the retrenchment of a breach, as by the Duke of Guise at Metz, in 1552, and by Montluc at Sienna, in 1555. The latter has described his arrangements in his commentaries.[6] Instead of disputing the breach, he made up his mind to retrench at a good distance behind the wall, and to let the enemy enter freely after a mere show of resistance, so as to do them the more mischief. At night he put scouts out, fifty or sixty paces beyond the wall at likely points, "to discover if there were not three or four who came to view that place, and to observe if they did not lay their heads together to confer; for this is a certain sign that they came to view that place in order to the bringing up of artillery. To do which as it ought to be done, they ought to be no other than the master of the ordnance, the colonel or the camp-master of the infantry, or the engineer, the master carter, and a captain of pioneers, to the end that according to what shall be resolved upon by the master of the ordnance,

[6] The Commentaries of Messire Blaize de Montluc, Mareschal de France. (Translated by C. Cotton, London, 1674.)

the colonel, and the cannoneer, the master carter may also take notice which way he may bring up artillery to the place; and the cannoneer ought to show the captain of the pioneers what is to be done for the esplanade, or plaining of the way."

The besiegers three or four times changed their point of attack, finding that, owing to the scouts, it was known to the besieged as soon as decided on, and that within an hour a thousand or more were at work retrenching.

The point finally chosen by them was one at which the houses of the city came so close to the walls that more than a hundred houses had to be pulled down to make room for the retrenchment or retirade; a necessity much regretted by Montluc, "for it is to create so many enemies in our entrails, the poor citizen losing all patience to see his house pulled down before his eyes." The Siennese, however, "put themselves their own hands first to the work."

While the besiegers were making the mound (*terrass*) for their guns, the besieged worked at the retirade, which was placed about eighty paces behind the wall. It was not much above the height of a man, and behind it were to be posted musketeers and harquebusiers. Behind a traverse at one end of it were three culverins, and at the other end five culverins, loaded with chain, nails, and scrap-iron; and when the stormers had been received by a general fire of artillery and small shot they were to be charged on both flanks by troops of about eleven companies (scil. 2000 men) each. In these troops "there was not so much as one harquebus, but pikes, halberts, and two-hand swords (and of these but few), swords and targets, all arms proper for close fighting, and the most furious and killing weapons of all others; for to stand popping and pelting with these small shot is but so much time lost; a man must close, and rapple collar to collar, if he mean to rid any work, which the soldier will never do so long as he has his fire-arms in his hands, but will be always fighting at distance."

After two nights and a day the besiegers had twelve guns planted, about one hundred paces off, and began to batter within a foot or two of the bottom of the walls, "which they did to cut the wall by the bottom, making account the next day with the rest of the artillery in a short time to beat down the whole wall." But in the afternoon a demi-cannon placed in an outwork enfiladed their battery, dismounted six pieces, and drove the gunners away for the rest of that day. In the night the retirade was finished, and a man was sent over the wall to see what damage had been done to it. He reported that "they had cut above fourscore paces of the wall within a span or two of the bottom, and that he believed in a few hours they would have beaten it totally down." But the Imperialist general, finding that here, as elsewhere, the besieged had retrenched, and were ready to fight him within the town, made no further attempt to take it by force. He sent his heavy guns away, and trusted to famine, which after three months proved successful.

But in order to gain time for making retrenchments, it was important that the wall should resist breaching as long as possible. Backing it with earth, as was done at Padua, deadened the vibration, which compromised its stability more than the actual destruction caused by the shot; and this means of strengthening it was commonly adopted. It was essential that this backing should not be merely of loose earth, exerting a heavy outward thrust upon the wall, and it was carefully carried up in well-rammed layers of small thickness, with trunks of trees, or brush-wood, or sods, between them. This was termed rampairing (*remparer*) a wall, and hence the mass was called a rampart.

The wall became a revetment for the rampart, and various methods of construction were adopted for it in new works. Some engineers relied upon mere bulk of masonry, others more skilfully disposed their material in a comparatively thin face-wall with long counterforts, and others again had recourse to arched construction. The face of the wall was usually scarped or sloping at the lower part, and vertical at the upper part; but sometimes this was reversed, to make shots glance off, and to make escalade more difficult. Sometimes it was wholly vertical, or wholly inclined. The counterforts were commonly vertical, but occasionally, as in the works of Civita Vecchia, built by San Gallo in 1515, they were made horizontal.[7] Arches were not only turned between the counterforts, but the face-wall itself was sometimes built up of them, ring above ring.

The use of arches lent itself to the formation of vaulted chambers, available for shelter or for active defence, and this was especially appreciated by Albert Durer; though long before his time it had been not uncommon to form a gallery behind the wall at its base, to allow the besieged

[7] The French engineers, when altering the fortifications a few years ago, found these walls in excellent condition. The counterforts were five feet long, one and a half feet thick, and three or four feet apart; the face-wall, which was sloping, was three feet thick, and twenty-six feet high. (Villenoisy, " Essai historique sur la Fortification.")

to watch for the enemy's miner, and to countermine as soon as he was heard.

But whatever might be effected by skilfulness or solidity of construction in the wall itself or the rampart behind it, the provision of a screen in front of it was the best means of preserving it. To sink it in the ditch was the most obvious step. Earth was needed for rampart-making, and the ditch, which had hitherto been only an occasional feature in fortification, valuable chiefly as a hindrance to the bringing up of engines, became henceforward essential and characteristic. It began to be recognized that, in Machiavelli's words, "ditches are the first and the strongest defences of fortified places." But old walls could not be sunk out of sight, and even with new works there was great reluctance to sacrifice high command. To use some of the earth dug out from the ditches to form banks to shield the walls, was the only way to reconcile protection with command; and in Italy the raised glacis, with the covered way behind it, was early developed. It is to be found in some of the designs of Francesco di Giorgio Martini, of Sienna, who was the military architect of Duke Frederic of Urbino, and died at a very advanced age in 1506.[8]

Apart from the protection of the wall, the glacis and the covered way necessarily followed upon the deepening of the ditch and the revetment of the counterscarp. Counterscarps of masonry were not unknown in mediæval fortresses. They were provided, for instance, at Ferrara in 1395, and Christine de Pisan shows that their value as

[8] "Trattato di Architettura Civile et Militare," first printed at Turin in 1841. The Corps Papers of the Royal Engineers (1848-9) contained an account of this work, by Captain Tylden, R.E.

an obstacle was well understood.[9] But such counterscarps were rare; and as the ditch was seldom deep, it served as a covered way. Their more general adoption, as the ditch grew deeper, hindered any free ascent from the bottom of the ditch to the ground outside; and by the middle of the sixteenth century the importance of converting the patrol road which ran along the top into a covered way was generally recognized in Italy.

Tartaglia, who from his mode of speaking of it has been sometimes credited with the invention of the covered way, proposed to keep it low enough to hide horsemen.[1] Maggi recommended that it should be made eight to ten yards wide, with a wall outside of it four to seven feet high, serving as a retaining wall to the glacis, but rising slightly above it. He advised that the slope of the glacis should be prolonged till it was nine or ten feet below the natural surface of the ground, so that the enemy might find himself checked by a sudden drop within good harquebus range, and if he jumped down might not easily be able to get back again. Some such obstacle was commonly provided, and the glacis thus formed a continuous outwork, which was often spoken of as the counterscarp.[2]

---

[9] "Also ought the ditches to be so deep and so large that they be not of light filled by the enemies, and some ancients made them in old time past to be masoned as a wall upright at the without forth side, so that one might not descend himself adown therein, and yet with this they strak full thick all downward the wall with sharp hooks and pins of iron that men call caltraps, that let right sore them that go down." (Op. cit.)

[1] "Quesiti ed Inventioni diversi, 1554."

[2] So Vigenére, a French writer at the beginning of the seventeenth century, says: "The counterscarp, bank, or douve, for they

There was another mode of partially screening the main wall which was largely adopted north of the Alps, namely by *brayes* and *bulwarks* in the ditch. As early as 1474 we find mention of this combination in the case of the town of Neuss, which was fruitlessly besieged for a whole year by Charles the Bold. The brayes while adding a fresh obstacle to assault, at the same time furnished a lower tier of fire; and the latter seems to have been their earliest purpose. The lists or bayles of mediæval places formed a covered way and place of arms guarded by a wall or a stockade, either at the base of the walls, or beyond the moat. This sometimes amounted to a lower enceinte, the value of which became greater as the main enceinte showed itself less able to withstand the besieger's cannon. In deepening ditches it was not safe to excavate close up to the old walls: it was necessary for their stability to leave a wide berm, at all events. So the new ditch was usually carried in front of the lower line, whether this had already a ditch behind it or not. This lower line became a cover to the lower part of the walls, especially if it was banked up with some of the excavated earth, and the term "braye" (probably from *braccæ*, nether garments) seems to have been applied to it on that account. Where it was close to the walls, with no ditch intervening, the cover afforded was more apparent than real, and hence, it has been supposed, the term *fausse-braye*, which continued in use long after "braye" had

are three names for the same thing, is a terreplein running along on the outside of the ditch, rising gradually from some distance off up to its brink, so as to make the ditch deeper, and increase the difficulty of filling it up, and of hurting the wall within it." (*See* "Le passé et l'avenir de l'Artillerie," ii. 270).

become obsolete.[3] At Orange, Augsburg, and elsewhere well-flanked faussebrayes were formed, and sometimes, these seem to have developed gradually into the principal enceinte.

Bulwarks to cover the gates were, as already mentioned, the earliest provision made against cannon, and the same protection was also sometimes given to towers. These earthworks were often capable of obstinate defence, as in the case of the Boulevard des Tourelles at Orleans; and they were soon found also highly convenient for offence. The narrow curtains of the old walls afforded no room for artillery, and the towers themselves were not well suited for it. If the guns were placed on the platforms of the towers the space was very limited, and the vault sometimes gave way; if they were placed in the floors below, their service was hindered by the smoke. High mounds were made for them inside the walls where it could conveniently be done; but often the best place for them was in the bulwarks outside. Here they were able to flank the walls and give a reverse fire upon the breaches; and before long bulwarks began to be made with this special object, and placed at intervals suitable for it.

[3] Viollet le Duc. It seems more probable, however, that the term is really a perversion of "fosse braye," and merely refers to its position in the ditch. Castriotto, describing an arrangement he had seen in some French fortresses—a low detached wall in the ditch masking the base of the main wall—speaks of it as a Fosse Bree or Fossa Brea (Book ii. chap. 25). De Ville, after treating of ditches, begins his chapter on Faussebrayes: " Nous pouvions mettre icy devant le discours des Fossez d'autant que les Faussebrayes sont tousiours plus proches des murailles que les Faussez : mais parce qu'on les fait dans les Fossez, il m'a semblé estre necessaire d'en parler premièrement." (L. i. ch. 38).

It has been remarked that fire-arms showed their value in defence earlier than in attack, and engineers studied how to turn them to account before they studied how to guard against them. So in Martini's designs, which belong to the latter part of the fifteenth century, the flank defence of the walls is kept far more steadily in view than the screening of them from the enemy. Sometimes he adopts a star trace with towers at the salients; or an indented or serrated trace, which will allow of flank defence from the top of the wall. But more frequently he adopts a simple polygonal trace, and flanks the wall by low chambers at its base, sheltered from the besieger's cannon. These caponiers, or *capannati* as he calls them, were evidently developed out of buttresses, though in some cases they are detached from the walls, and their shape varies. They should be twelve or fourteen feet wide, he says, with loopholed walls, five or six feet thick, and flues to carry off the smoke. "The capannato should be contiguous or attached to the wall of the fortress with a narrow covered entrance from the interior of the work, arranged so that when the capannato shall be lost, the fortress shall not be taken also."

Casemates, as small caponiers of this kind were then commonly called, played an important part in several sieges of the sixteenth century, and were used by many engineers. But the name of Albert Durer is especially connected with them. Combining them with his vaulted walls, he provided for his ditch a grazing artillery defence both direct and flanking; and he stands quite alone in the development he gives to them, in the solidity and convenience of their construction, and above all in the careful arrangements made to clear them of smoke. "No shot," he says, "should be fired without the muzzle of the gun being made to project beyond the embrasure, otherwise a thick smoke is at once forced backwards and causes great inconvenience."[4] Accordingly the wall is hollowed out to allow the guns to be run forward sufficiently. But he also provides large smoke flues, three feet in diameter, to ventilate the casemates, and his caponiers have a central passage left open above. His projects, however, were on too gigantic and costly a scale to be ever carried out, and for a time fortification developed in a different direction. The Italian engineers objected to his caponiers, that the besieger would have little difficulty in sapping across the ditch, if he had only to screen himself from their grazing fire. To obtain a more plunging flank fire, they set to work upon the bulwarks, and transformed them gradually into the modern angular bastion. Of this form some of the bastions built by San Michele at Verona, in the second quarter of the sixteenth century, are probably the oldest surviving examples. But it has been found in a sketch of the works constructed by the Florentines at Pisa, in 1509-11, and also in some designs which, from similarity of style and paper, are attributed to Martini. It cannot be assigned as an invention to any one man or one place, but was merely one among many varieties of form simultaneously tried. The word "bastion," like the word "rampart," denoted originally not the shape or function of a work, but the character of its construction, that it

---

[4] His work, "Etliche Unterricht zu Befestigung der Stadt, Schloss und Flecken," was published at Nuremberg in 1527, and a Latin translation of it at Paris a few years afterwards. A French translation, by Captain Ratheau, was published in 1875.

was well consolidated; and old French writers seem to use "bastille," "bastillon," and "boulevard" indiscriminately. The Italian equivalent of bulwark, "baluardo," was the term commonly used in Italy throughout the sixteenth century, though occasionally "puntone" was employed to denote angularity of form.

When bulwarks were thrown out beyond the main ditch, the Italians styled them "rivellini" or scoutworks, and the term *ravelin*, derived from this, was at first applied generally to outworks; though, like the more northern term, "halfmoon" or "demi-lune," which seems to have been used with equal looseness, it has gradually been confined to the principal outwork which covers the middle of a front. In the same way the term "bastion" was appropriated by degrees to bulwarks placed near the main wall, and connected with it by flanks defending it. These flanks, or at all events their upper tiers, for there were usually two tiers at least, were retired so as to be well sheltered from outside view by the extremities of the faces. So well sheltered, some engineers complained, that they themselves could see nothing; but the value of one or two hidden pieces which opened fire unexpectedly at the decisive moment when the enemy was checked upon the breach, was often illustrated. This orillon construction was probably suggested by the earlier application of bulwarks to towers, the lower stories of which continued to furnish a flank fire along the curtains. As a rule the flanks served only for the defence of the curtains. The faces of the bastions were flanked either from *cavaliers* raised upon the curtains, or from *platforms* projecting in front of them. But there was a strong objection to acute salients, and it was often thought better to blunt them even at the sacrifice of flank defence; or some self-flanking trace was adopted, as at Augsburg and by the engineers of Henry VIII. There was not much concern about that complete avoidance of dead angles in which has since been found the *raison d'être* of the bastioned trace. Maggi, one of the joint authors of the best Italian work on fortification at that time,[5] was at great pains to bring together and compare the opinions of different engineers of reputation. Some contended that the curtains should be between 300 and 400 yards long, so that the flanks might be out of range of one another, others preferred about 200 yards, notwithstanding the cost of the increased number of bastions. The length of the faces of the bastions should be forty to fifty yards, and of the flanks thirty to forty yards, the latter perpendicular to the curtains. The height of the wall should be thirty to forty feet, the faces of the bastions being about three feet higher than the flanks and curtain. On the flanks the floor of the upper tier might be about twenty-five feet, and of the lower tier about eleven feet above the bottom of the ditch; and below these there might be casemates for a grazing fire of harquebuses, though the smoke would be found inconvenient. Opinion ranged from twelve yards to thirty yards for the width of the ditch, and from ten feet to twenty feet for its depth.

The reconstruction or amendment of fortified places went on actively all over Western Europe during the first half of the sixteenth century. The feudal castles had received a death-blow from the new artillery, as Franz von Sickingen found at Landstuhl in 1523; and they could not be effectually remodelled without demanding larger

[5] " Delle Fortificatione delle Città," di M. Girolamo Maggi e del Capitan Iacomo Castriotto, Venice, 1564.

garrisons than the barons could muster, so that private war perforce declined. But the wealthier cities, Augsburg, Nuremberg, Frankfort, Lubeck, Antwerp, and others, hastened to reinforce their walls by the new methods. In France it became a national question. Taxes were levied to be exclusively applied to works of defence, controllers were appointed to see that the money was properly spent, and, as elsewhere, engineers were invited from Italy to direct the works. In England, Henry VIII., who entered into matters of military engineering, "not with the condescending incapacity of a royal amateur, but with thorough workmanlike understanding,"[6] studded the coast with castles of a curious transition type, which are still to be seen at Deal, Sandgate, or Portland. These were merely coast-forts, to deal with shipping or with a hurried land-attack; but Calais had to be guarded against a prolonged siege, and extensive works were carried out there, which caused it to be reputed one of the strongest fortresses in Europe. In 1540 nearly a thousand men were employed there, and almost as many at its outwork, Guisnes, and the monthly expenditure was nearly 3000*l*. Bulwarks and brayes were made to cover the towers and curtains, and the latter were cut down and made solid, and otherwise adapted for artillery, the embrasures being shaped "as the king's grace hath devised."[7] Batardeaux were provided across the ditch and across the harbour to hold the flood-waters, and surround the place with an inundation. But after Henry's death the defences were neglected, and began to fall into decay. In 1557 the

[6] Froude, "History of England."
[7] The "device" or project for the alterations at Calais is given, together with a plan made about 1540, in Nichol's "Chronicle of Calais" (published by the Camden Society in 1846).

Venetian ambassador, after speaking of the vital importance of Calais to the English, inasmuch as, "if they were deprived of it, they would not only be shut out from the continent, but also from the commerce and intercourse of the world," hinted doubts of its strength;[8] and in the following year the Duke of Guise showed that these doubts were well-founded. His attack was made from the north, where the old walls had not been reinforced, the inundation was not too deep for his men to wade through, and the garrison, few in numbers and in bad heart, offered little resistance. Guisnes was attacked a fortnight afterwards, and made a better fight. The story of its defence has been well told by the son of the governor, Lord Grey,[9] and the obstinate struggle for one of the new bulwarks is worth quoting. It brings out very strikingly the value of hidden flanks.

"The town was a large thing, and my lord unable anything like to man the same; wherefore, about midnight, setting all the houses on fire, he drew all the soldiers into the castle. Till the Monday next we kept them such play with great ordnance and often sallies as scarcely to cast their trenches, much less to plant any battery they were able. On the said Monday morning, by the break of day they had laid two batteries to the Mary bulwark, thirteen cannons in the one, and nine in the other, with which they plied it so well that by noon they had not only dismounted our counterbattery, but also clean cut away the hoop of brick of the whole forefront of our bul-

---

[8] "Chronicle of Calais."

[9] "A Commentary of the Services and Charges that my Lord my Father was employed in whilst he lived." By Arthur, Lord Grey of Wilton. Published by the Camden Society, 1847. Lord Grey's narrative was used by Holinshed.

wark, wherewith the filling being but of late digged earth like sand did crumble away; which the enemy finding, about two of the clock of the same afternoon sent a forty or fifty forlorn boys with swords and roundels to view and assay the breach. The ditch at the place before the battery was not twenty-four feet broad, now assuredly not a dozen, nor in depth above a man's knees; wherefore with small ado they came to the breach and with as little pain ran up the same, the climb was so easy; from whence, having discharged certain pistols upon us, and received a few pushes of the pike, they retired; and making report (of like) of the easiness of the breach, straight a band or two of Gascons (as it was thought) throw themselves into the ditch and up they come. Then a little more earnestly we leaned to our tackling, our flankers walked, our pikes, our collars, our pots of wild-fire were lent them, the harquebus saluted them, so as jolly Mr. Gascon was sent down with more haste than he came up with good speed; and so ended Monday's work, saving that upon the retire from the assault they gave us six or seven such terrible tires of battery as took clean away from us the top of our vammures [1] and maunds, leaving us all open to the cannon's mouth; whereby surely but for night that came on we had been enforced to have abandoned the place."

It was necessary during the night "for the winning of a new vammure to entrench within the bulwark six feet in depth and nine in thickness, which marvellously did straiten the piece, the same being of no great largeness before."

"By the next day, being Tuesday, they had planted two batteries more, the one in the market-place of the town, of six cannons, to beat a curtain of the body of the castle; the other upon the rampart of the town, to beat

---

[1] *Avantmur*, or parapet.

the cat, and a flanker of the barbican, which two commanded one side of Mary bulwark. This morning the enemy most bestowed in playing at our flankers, which the day before they had felt, and indeed well-nigh took every one from us, saving that of the cat, which lay high and somewhat secret, and another at the end of a braye, by the gate on the other side of the bulwark; all the rest, as those of the Garden bulwark, (which chiefly beheld the main breach), the barbican and the keep were quite bereaved us, and besides they continually entertained the breach with ten or eleven tires the hour. In the afternoon, about the same hour as the day before, a regiment of Switzers with certain bands of French, approached the dyke as if presently they would have given the assault; but there did stay, sending to the breach only a company or two, seeking thereby to have discovered what flankers yet were left us, wherein they were prevented, my lord having before warned the gunners not to disclose them but upon extremity; and thus after an hour's play with the harquebus only, and a light offer or two of approach, this people retired them, and gave the cannon place again, which by night had driven us anew to become moldwarps." Next day, the enemy's cannon having played all the morning, "and well searched as they thought, every corner that *casematti* might lurk in, about one of the clock we might descry the trench before the breach to be stuffed with ensigns. My lord straight expecting that that followed, gave word straight to every place to stand on their guard, encouraging every man to continue in their well begun endeavours. A tower that was called Webb's tower and yet standing, that flanked one side of the beaten bulwark, he stuffed with a twenty of the best shot with curriers. These things no sooner thus ordered, but that eight or nine ensigns of Switzers and a three of Gascons do present themselves upon the counterscarp; and without stay the Gascons fly into the ditch, the breach they run up, our harquebusry receiveth them, they two for one requite us. The top of our vammure, or rather trench, they approach, the pike is offered, to hand blows it cometh; then the Switzer with a stately leisure steps into the ditch, close together marcheth up the breach; the fight warmeth, the breaches all covered with the enemy. The small shot in Webb's tower began now their parts, no bullet that went in vain. On the other side again, twenty of the Spaniards on the inside of the brayes, had laid themselves close till this heat of the assault, and then showing themselves did no less gall the enemy than the tower; thus went it, no lustilier assailed than gallantly defended. At last after an hour's fight and more, the governors without, finding the great slaughter that theirs went to, and small avail, and perceiving the two little *casematti* of tower and brayes to be the chiefest annoyances, did cause a retire to be sounded, and withal three or four of the cannons in the market-place to be turned upon Webb's tower; the which at two tires brought clean down the same upon the poor soldiers' heads, wherein two or three were slain outright, others hurt to death, and they that scaped best so maimed or bruised that they were no more able to serve.

"The enemy this while having breathed, and a brace of hundred shot put forth only to attend on the few Spaniards that kept the corner of the brayes, the assault afresh is begun, and their beaten bands with new companies relieved. My lord also sent into the bulwark

200 fresh men. Now grew the fight heavy upon us, all our defence resting in the pike and bill, our chiefest flankers being gone, our places to bestow shot in taken from us, our fireworks in manner spent, the Spanish shot on the other side so overlaid as not one of them but was either slain or marred ere a quarter of the assault was past. The easiness of the fight thus alluring the enemy, unappointed companies flew to the breach and courage was on every side with them. My lord perceiving the extremity, sent to the two fore-named flankers that they should no longer spare; they straight played: the ditches and breach being covered with men, what havoc they made it is not hard to guess. These unlooked for guests made the enemy that was coming to pause, and the others already come to repent their haste. Three or four bouts of these salutations began to clear well the breach, though the ditch grew the fuller. The night at last parted with no great triumph of either winning, for as we went not scotfree, so surely no small number of their carcases took up their lodging that night in the ditch."

During the night the noise of workmen was heard in the ditch; "at the last with cressets it was espied that they were making a bridge. The morning come, we see the same finished, empty casks with ropes fastened together, and sawed boards laid thereon. They spent all day till it was full three of the clock in battery, and beating at our two last flankers which at last they won from us and the gunners of either slain." The governor then decided to abandon the bulwark as soon as a mine had been got ready there to blow up the enemy; but before this could be done, a new and more furious assault was made, and after a hard struggle the defenders were driven out of it into

INTRODUCTION. 35

the castle. The soldiers in the other bulwarks and in the base court followed them, and forced the governor, much against his will, to surrender without further fighting.

Throughout the sixteenth century Italian engineers were in general request throughout Europe, and Italian works on fortification poured from the press. Authors vied with one another in displaying their ingenuity, and endeavoured to furnish cut-and-dried solutions for all the various problems that might present themselves in practice. Marchi, who presented a copy of his work to Philip II. at Greenwich, in 1556, though it was not published till the end of the century,[2] prided himself on having contrived 161 systems. Zanchi complains that "sundry being able to make the draught or portraiture of a thing give themselves to the making of plates of fortresses without number, not understanding nor knowing how to proceed nor to discourse from the foundations of the matter with lively reasons;"[3] and this complaint of mere draughtsmen-engineers meets us again and again.

"There is not a painter, carver, mason, joiner, carpenter, architect, or in fact any class of men whatever, that have not employed their pencil on it; as if the thing lay in representation, and in knowing how to draw a line straight or curved with rule and compass, and inventing the design of a fortress out of one's own head, and were

[2] "Della Architettura Militare." Brescia, 1599.

[3] "Key of the Treasury." This is a MS. work in the British Museum, dated 1559, and based upon the treatise of Zanchi of Pesaro, "Del modo di fortificar le Città" (published at Venice in 1554). It was probably the earliest English statement of the principles of the Italian Engineers. An account of it, with extracts, is given in Part II. of the "Woolwich Text-book of Fortification."

not an art slowly acquired by a large experience of sieges, both in attack and defence." [4]

Paper systems, whether good or bad, could be applied only partially and imperfectly in remodelling the defences of towns, whose existing walls had to be turned to account, and whose accidents of shape and site necessarily caused irregularities. But the comparative weakness of such patchwork made it the more important to provide an inner line of defence, or keep; and the old castles, so far from being adequate to this, were often—as in the case of Calais—the earliest works to fall. Citadels were therefore made in which the garrisons might hold out after the town had been taken, and which would serve the further, and often more essential, function of keeping the burghers in check. In the construction of these citadels—small, strong, wholly new, and upon chosen ground—engineers had freer scope. At the beginning of the century, when the Florentines recovered Pisa, their first care was to make a new citadels and in 1567, when Alva came to the Netherlands, he brought with him a noted engineer, Paciotto d'Urbino, who had already built citadels at Turin and Cambrai, and was at once sent to build one at Antwerp. Two thousand workmen were employed upon it daily, and it was finished in little more than a year, at a great cost. It was reckoned the masterpiece of the age, and Strada says that Paciotto "got himself a great name by it, being from thence called the inventor of modern fortification." A recent writer [5] has endorsed the judgment of that

[4] Vigenére (1615), quoted in the "Études sur le passé et l'avenir de l'Artillerie," ii. 26 4.
[5] Colonel Cosseron de Villenoisy.

day, on the ground that he was the first engineer who constructed a bastioned front with its several parts in due proportion. The flanks, as usual then and for some time afterwards, were made in two tiers and at right angles to the curtain, but in other respects the fronts agreed with modern rules. Inside, there was barrack-room and store-room for 5000 men.[6] But this sumptuous fortifying, however satisfactory to engineers, did not escape the criticisms of men familiar with war. La Noue "of the iron arm," one of the Huguenot leaders, and one of the best soldiers of France, gives expression to them.[7] "The Italians deserve the commendation of being the first inventors of divers sorts of gallant fortifications, which since they have reduced into such an art as hath been esteemed honourable, neither hath it been of less profit to those that have dealt therein. And peradventure this last point hath partly been the occasion that they have persuaded princes that such and so many things were requisite to bring a piece of work to perfection and worthy of them . . . The first place that here I will bring to view shall be the citadel of Antwerp, wherein we may say that nothing hath been forgotten, either in wealth, diligence, invention or plenty of stuff; so as in all Christendom a goodlier piece of work for

[6] After an eventful history of 300 years, Paciotto's citadel has lately been superseded and swept away. Paciotto himself has been confounded by many writers with a Spaniard, Colonel Pacheco, who was hanged by the men of Flushing in 1572; he died at Urbino in 1591.

[7] "Political and Military Discourses, 1587." The quotation is from an English translation, published in the same year (p. 215). See also Busca ("Della espugnatione et difesa delle Fortezze." Turin, 1585.)

fortification hath no man seen. But, on the other side, if we consider that the building thereof cost 1,400,000 florins, and yet, had it been assaulted, would not peradventure have held out much better than Oudenarde or Maestricht, which were fortified but with earth, it will make men somewhat curious to examine these matters more exactly . . . . The engineers shall say that notwithstanding men fortify but with earth without any of their supporters of stone or brick (which are no less beautiful than necessary) yet still they follow their precepts. Whereto I answer that in many things men may help themselves therewith : howbeit they are rather to stick to new experiences which have taught very good kinds of fitting and defending themselves.

"The first is the same that I have already mentioned, namely, fortification with earth: which costs ten times less than great masonry, and is never a whit worse. . . .

"The second thing which experience hath made many to allow of, is to loosen the bastions from the curtains, yea and to carry them without the ditch ; for although they be not defended with the artillery from any low casemates, yet do the harquebusery sufficiently shield them from the curtains which is a continual annoyance that cannot be taken away, where [as] the flanks of the bastions may be pierced or broken when the shoulders are weak. Also if one of those ravelins that I speak of should chance to be taken, yet is not the place therefore so lost, but that the enemy may very well be put back, where contrariwise it is a necessary consequence to those that have joined them to the ramparts.

"The third is the use of intrenching (*scil.* retrenching), which is a marvellous profitable remedy though smally practised in times past, but in our civil wars men have learnt to use it very well. Though they be weak and but ill made, yet do they preserve from being forced on a sudden, and procure some reasonable composition. But if they be large and well made, either they wholly preserve or at the least do give a moment's respite (which is a sovereign purchase to the besieged, when the enemy must win it by little and little), during which time they may light upon some other favourable accident for themselves."

In La Noue's remarks, and especially in what he says about retrenchments, we see the influence of the famous defence of Rochelle, in 1572, in which for a time he bore a part. There, behind old-fashioned walls, reinforced at some points by detached bastions, roomy but ill flanked, the Huguenot citizens kept at bay the best troops of France for more than four months, repulsing eight assaults, and costing the besiegers a loss of 20,000 men.[3]

The desperate courage of the inhabitants and the lukewarmness of the besiegers, of whom a large number were Protestants, was no doubt the chief cause of so successful a defence of a town by no means strong. The leisure allowed to the citizens for preparation, the want of subordination among the chiefs of the Catholics, and the indecision of their commander, were further causes; but this, nevertheless, like other sieges of that time, shows how favourable the conditions still were for the near defence. Before the breach could be made by the besieger, a retrenchment could be made by the garrison; mining was a tedious and an uncertain resource, a matter of

[3] Vide "Spectateur Militaire," 1848.

knack rather than rule; and so long as the "flankers" were undestroyed good troops could hold a breach against an army. The citizens had built "casemates" in the ditch where the flank defence was worst, and both their merits and defects were plainly shown in this siege. It was very difficult to bring artillery to bear on them, and, however the sap might give cover against them across the ditch, they were able to take the stormers in reverse as they mounted the breach; on the other hand, the *débris* from the walls and ramparts was apt to mask their loop-holes, and the besieger might make himself master of them either by sudden assault, or by the slower but surer method of sapping up to them. Especially where, as in this case, there was no gallery of communication, but the entrance to them was direct from the ditch, only good soldiers could be trusted to hold them.

While the siege of Rochelle was going on, the defensive force of an active and stout-hearted garrison, even though ill-trained, was being still more strikingly illustrated elsewhere. At Haarlem the long continuance of the defence could not be attributed to any half-heartedness on the part of the besiegers, or any weakness or division among their leaders. The troops were reputed the best soldiers in Europe, and they were only too eager for their prey; and they were led by Alva's son under the guidance of Alva himself.

They made the mistake, however, at first of despising their enemy. To save shifting their camp, they chose a point of attack which was not the most favourable. The artillery ammunition ran short, and so they tried an assault before a good breach had been made, and when the trenches had only been brought up to the batteries.

The assault failed; they had to wait for more ammunition and to push their trench right up to the ditch. The trench was made in a novel way by Bartolommeo Campi, the engineer. It was carried straight to the front, but bridged, or blinded, at intervals by joists, which were supported on uprights and covered with sandbags. These formed overhead traverses which prevented men in the trench from being seen from the front. To give more fire, and to hold more men than there was room for in the trench, branches were run out from it to right and left. At length, at the end of a month, the besiegers, having filled up the wet ditch, succeeded in getting possession of the ravelin which covered the gate they were attacking; and made in it a sand-bag battery for two guns, to search out the flanks of the main wall. But during this month the citizens had been so busy that "their town was three times stronger than the first hour the enemy encamped before it."[9] Behind the breach in the main wall they made a semi-circular retrenchment, about 200 feet long, joined to the wall at each end, and with a wide wet ditch in its front. Constructed of old ships, earth, old houses pulled down, and anything else that was suitable for filling, it had an open parapet for musketeers above, and gun-rooms underneath.[1]

Receiving supplies and reinforcements from outside, they were able to hold out for six months longer, making frequent sorties, repulsing assault after assault, and meeting the besiegers' mines by countermines, which made the breach so steep that it could not be climbed.

[9] Williams, "Actions of the Low Countries," 1618.
[1] Speckle, "Architectura von Vestungen," 1587.

But after the winter was over the investment became more effectual, and famine at last obliged the city to surrender.

The resistance made by Haarlem, and by other places much less considerable, taught the Spaniards caution. They found that even against mere citizens the place was not won because the breach was made; and they learnt to temper their valour by discretion, as we see in the picture drawn of them by Sir Roger Williams.

"When they assiege a place they encamp at the first out of danger of the enemy's artillery: before they make any approaches they do what they can to make sure either with forts or trenches all the passages, I mean the coming in as well to their camp as to the town or place assieged. If they can make a bridge to pass over horse and foot with all necessaries from one side unto the other; if they can they will not fail to make ways, round about the place assieged to march with horse and foot.... If they think the place assieged too well manned or the seat such by nature, that batteries can do no good, they will block it up with forts in such sort that half their army will be sufficient to assiege it. They will be sure to place the rest in the best quarters for victuals and forage, although it be three days' journey from the place assieged.... By these means they relieve their wearied troops with fresh at their pleasure. If there be troops making head to levy their siege, they will join closer together as occasion present. If they batter, they approach carefully with trenches afar off, spare neither pioneers nor cost to save their soldiers. Before they place their battery they mount culverins and other pieces to beat the flanks and defences: if there be not high grounds advantageous to do it, they will be sure whatsoever it costs to raise mounds for the purpose. If the bulwarks be such that the flanks cannot be taken away with their pieces, they will lay battery at once both to bulwark and curtain; for the soldiers may lawfully refuse to assault until the flanks be taken away; neither will the chiefs offer it for some of them must lead them." [2]

Mendoza, the historian of the war in the Low Countries, who was himself present at the siege of Haarlem, afterwards wrote a treatise for the Prince of Castille, in which he describes how the attack of a fortress should be conducted.[3] After speaking of the ordinary winding trench, and of the importance of tracing it so that none of its windings can be enfiladed, and making it wide and deep, he mentions approvingly the direct traversed trench used at Haarlem. As regards batteries, the sites chosen for them should be convenient for the service, defence, and withdrawal of the guns, and suitable for breaching the wall at a point easily accessible to the stormers. "Some hold that battery for the best when the pieces may be four score or one hundred paces from the wall, their fury being nothing so great when they stand one hundred and fifty or two hundred off.... For which cause others are of opinion that the pieces were best to be placed, if it were possible, upon the very brim of the ditch." But if they are too near the gunners will be annoyed by the harquebusiers of the place, and the garrison will be likely to sally out and spike the guns before the guard can be reinforced. The pieces " being placed with their beds, which

[2] Williams, "Discourse of War." London, 1590.
[3] "Theorique and Practise of Warre," written in 1594, and translated by Sir E. Hoby in 1597.

are made of timber planks and hurdles, they cover them with gabions and ditches." The heavy guns should fire by volleys, so that the shock to the wall may be the greater, and the lighter pieces should keep the enemy in check between the volleys, and cut away the shaken masonry. If the wall is unbacked by earth, it should be battered aslant, as otherwise the shots pierce without shaking it. Busća, an Italian engineer of that day, prescribed this even for earth-backed walls, and in Sully's artillery instructions it was laid down that to breach a curtain there should be eight cannon firing perpendicularly to it, and on each side of them four lighter pieces firing a little obliquely. Busca recommended that in breaching a wall a horizontal cut should be first made at about one-third or one-fourth of its height, but that earthworks should be cut down by degrees from the top.

At night, Mendoza says, the guns should be fired singly, at intervals, to hinder the enemy from repairing the breach ; and the ditch may be reconnoitred, to ascertain whether there are any low flanking casemates in it, if it is dry, or if wet, to devise how to drain it, or dam it with faggots or earth. For a deep wet ditch bridges must be got ready, made of barrels, barks and ships' masts ; but the recollection of Haarlem made him very averse to assaults under such conditions. If it is necessary to sap up to the walls, the soldiers, as they work, protect themselves with "blinds" (post and rail frames with short upright fascines secured to them), and cover these on the outside with hides to prevent their being set on fire. In speaking of mining, he mentions the ancient method as still in use : sometimes, he says, " they put posts, digging under the foundations, and when they see that they only support the wall, they anoint them with tallow and pitch that they may burn the better, putting powder about them, and a quantity of straw and wood, to which they put fire when the men are ready to assault upon the falling of the wall."

The Dutch war of independence was a war of sieges, and gave abundant opportunities for improving the art of attack. In Maurice of Nassau, too, it produced a leader to make the most of those opportunities. Only eighteen when he was appointed captain-general of Holland and Zealand, he showed at once that he appreciated the change then coming over warfare, and the full potency of science and system. He contributed to that gradual supersession of the pike by the musket, which had been initiated by the great Spanish captains before him, and was afterwards continued by Gustavus and consummated by Vauban. In each company of foot he had sixty-four "shot" to thirty pikemen or halberdiers, at a time when it was commonly thought that the pikemen ought to be the more numerous.[4] He saw, too, how large was the part to be played by field intrenchments, whether as a substitute for the body armour, which was now being discarded owing to the increased power of firearms, or as an obstacle to serve instead of pikes to keep away the horse. To turn them to full account, they must not be left, as hitherto, to bodies of civilian pioneers, but must be undertaken by the troops themselves; and he showed that the contemptuous aversion of soldiers to spade-work was to be conquered by good extra pay. But competent direction

---

[4] Barry, an Irishman who served under Spinola, puts the pikes at fifty-five per cent. of the infantry.

was no less necessary than willing labour; and for the first time, under his orders, engineers were required to go through a regular course of instruction, drawn up by Simon Stevin at Leyden. He "loved mathematicians and engineers very well," writes his biographer,[5] "but there was nobody could teach the prince in that science, he having contrived several fine inventions for the passage of rivers and siege of places, so that in his age he served for a pattern to engineers as well as captains." But it is not as the rival of Pompey Targone, the author of so many adventurous schemes at the siege of Ostend, that Prince Maurice is now remembered. His fame rests upon the thoroughness and method of his work in sieges, as in everything else that he undertook: the foresight, judgment, and unsparing industry, that left no room for failure, and made besieged garrisons hopeless of escape. It was in his lines of circumvallation that these characteristics were perhaps most marked. His sieges were slow as compared with those of later days: he was three months before Gertruydenberg, and two months before Grave; but he always took care that he should not be interrupted in them. At Grave the parapets of his lines were "such as had never been seen in any army, and a ditch, a pike and a half deep, the parapets flanking each other thoroughly, with many redoubts or small forts scattered all along within musket-range of one another, so that there were fully seventy such forts in the whole circuit."[6] Before his death he had taught his enemies to imitate if not to outdo him. His great rival, Spinola, when blockading Breda in 1624, made lines of circumvallation of which the

[5] Aubrey Du Maurier.   [6] Hondius.

total length was reckoned at fifty miles, in addition to lines of contravallation, and forts, bastioned and demi-bastioned, redoubts and batteries, to the number of nearly 200. Breda had been fortified by Maurice with great care, and was especially cherished by him, being an old possession of the Nassau princes, and one of the earliest trophies of his career. But he tried in vain to relieve it after the lines were made. His health broke down under the disappointment, and he died, asking with his last breath, "Is Breda rendered or freed?"

The two opponents, the Italian and the German, offered a striking contrast. "The Marquess was very lean, the Prince very fat, and their tempers very different, the one being dry and choleric, the other plump and sanguine." "The prince, though he was very vigilant and laborious, yet had so great a quietness of mind that so soon as ever he was in bed, and his head laid upon the pillow, he fell into so sound a sleep that it was a difficult matter to wake him."[7]

Spinola's leading characteristic, on the other hand, was inexhaustible energy. "Being awakened he looked with such lively eyes, he listened with so attentive a mind, that he seemed not to have slept at all: and he as soon recovered his sleep as he was awaked, so securely slept he, all his cares being so well husbanded . . . . Touching the cruelty of the season, and the weather, and whether it rained or snowed, or freezed, or blew, or whether it were evening or midnight, he cared not. . . . He made no account of his meat, of his rest, nor of his own body. To provide for, to consult, write, hear, command, to go

---

[7] Aubrey Du Maurier.

about the camps, was like unto a daily pastime. Never did he pretend any excuse for his weariness; never forbid access unto him." [8]

Ample details of the investment and other siege-works, as matured under Maurice and his brother Frederic Henry, are given by Marollois, Freitag, and the other Dutch writers of that time. The first thing, when the lines were finished, was to determine the point of attack. This, says Marollois, will usually be a bastion, for the curtain, being defended on either hand is much more difficult to attack, especially if it is covered by an outwork; but with old-fashioned places, having long fronts and small bastions, the curtains were often preferred. The trenches should start from beyond musket-range of the place, or about 300 yards off, and be traced in zigzags, with an average length of about 120 yards for each branch. To support the workmen and repel sorties, redoubts, about twenty yards square, should be made at the angles of the zigzags, or in the course of the branches, when these were long. The Spaniards, instead of making redoubts, often followed the plan adopted by Montluc at Thionville,[9] or carried onward the hinder branch of a zigzag to form a flank for the one beyond it. Occasionally lateral trenches were

[8] "The Siege of Breda," by Captain G. Barry. Louvain, 1627.

[9] "At every twenty paces I made a back corner or return, winding sometimes to the left hand, and sometimes to the right, which I made so large that there was room for twelve or fifteen soldiers with their harquebuses and halberts; and this I did to the end that should the enemy gain the head of the trench, and should leap into it, those in the back corner might fight them, they being much more masters of the trench than they who were in the straight line."—Commentaries.

thrown out (or demi-parallels, as they would now be called), and redoubts placed at the end of them. According to Marollois' disposition, as the approaches neared the front attacked, they should spread outward, and then interlace, so as to furnish a wide front for the musketeers to support the further advance up the glacis, and the passage of the ditch.

When the trenches had been brought as near as they could be with any safety to the counterscarp, the "sap" must be employed, and should advance direct upon the bastion. The trenches were made only three feet deep, but the sap was sometimes six feet. According to Vauban's definition, "the sap is a kind of gallery completely sunk in the ground, by means of which one advances towards any works of the enemy. The uncovered sap, or half-sap, is a sap which is only half sunk, the rest remaining uncovered."[1] In sapping, "a man kneeling upon his knees digs to get into the ground, and casts up the earth before him on both sides with a short spade towards that part of the fortress, till he hath digged three foot into the ground, and that he is covered with the earth, casting always the earth like a mole before him towards the town."[2] Others followed to widen and deepen the trench. It was sometimes roofed, but more often it was screened from the enemy's view by the overhead traverses which have been already mentioned. "Chandeliers," as they were called, consisting of two pointed uprights connected by a groundsill, were placed at intervals on the side parapets of the trench, parallel to

[1] "Mémoire" of 1669.
[2] Hexham's "Principles of the Art Military" (London, 1639); a translation of Freitag as regards fortification.

one another, and fascines were then laid across the trench, between the uprights ; or instead of these, gabions on planks were sometimes used.

But the nature of the ground in some cases would not allow of deep trenches or even of trenches at all. At Bois le Duc, in 1629, the besiegers pushed an approach direct upon the city, across a morass " which was overflowed with water at some places a man's height." They made a causeway of earth and brushwood, rising one foot above the water, and formed parapets on each side, with an interval of twelve feet between them. To cover this roadway from the front, they built up traverses, eight or ten feet apart, and attached alternately to one parapet or the other, just as in the double sap of the present day. Parapets and traverses were alike made proof against artillery.[3]

At the great siege of Ostend, which lasted from 1601 to 1604, the difficulties which the marvellous energy of Spinola at last overcame, called for many new expedients. Never perhaps were approaches carried forward under more adverse conditions. The besiegers on the west side of the town had to work over sands covered by each tide, and to cross the old channel, in which there was always some water. Trenches were out of the question, so they at first formed parapets with gabions; but this could only be done by night because of the musketry fire of the garrison. As the artillery fire had been subdued, and only a bullet-proof parapet was needed, they made use of fascines instead of gabions, either securing them upright

[3] Prempart, "Siege of the Busse." Amsterdam, 1630. See also Occasional Papers of the R.E. Institute. Vol. VI. Paper No. 2. Plate 2.

to blind-frames, or piling them horizontally between chandeliers. These brushwood breastworks could be put together out of range, and floated into position on casks at high tide. But portions of them were burnt by the besieged with fireballs, and portions were swept away by the sea during gales. In any case the progress was slow and the loss heavy. "It was well," says one of Spinola's engineers,[4] "if half the working party came back unwounded when their task was done. Usually one man undertook to execute a length of blind for a certain sum, and then sought out comrades, making as much as he could by it, and paying them in proportion to the danger; and at times they were so handled that the greater part of them were dead or wounded, and others had to be found to finish the work they had begun. By such accidents it sometimes took from four to six whole nights to execute four hours' work."

How completely war was at that time a trade is curiously illustrated by the matter of working pay. As there were no regular corps of sappers, siege-works became a commercial transaction. A general having reached a suitable point, bargained with his workmen to make him a battery. The sap was executed by trained men who were highly paid for it, usually at two or three florins per yard, according to Marollois. But the rate depended on the danger, and he remarks that the besieger has to decide whether to be economical of money, or of time. Hexham recommends the besieged to use their cannons freely against the approaches, for then the enemy must either raise the siege or it will cost them dear, "because no

---

[4] Giustiniani, "Delle Guerre di Fiandra," 1609.

men will undertake the work but such as will be soundly paid for it."

The passage of a wet ditch was made with a wooden gallery, about seven feet wide and eight feet high, carried upon a causeway of fascines, which was formed as the gallery advanced. The top and one side of the gallery, sometimes indeed both sides, were covered with earth, so that it might not be set on fire.

Breaches, in the early part of the seventeenth century, were usually formed by mines, after the passage of the ditch had been completed, for artillery had little effect on the earthen slopes of the Dutch fortresses. Of Sandhill, one of the works of Ostend, we are told that it was rather fortified than injured by the severe battering it endured; "for by this time it was so thick stuck with bullets that the ordnance could scarcely shoot without a tautology, and hitting its former bullets, which, like an iron wall, made the latter fly in pieces up into the air."[5]

The part of artillery was to subdue the fire of the garrison, to assist in the repulse of sorties, and to prepare the way for assaults. The guns were usually raised on banks or mounts, three or four feet above the ground, so that they could fire over the trenches; and sometimes, as at Ostend, very high cavaliers were made by months of labour in order to obtain a plunging fire. The batteries were commonly open, but were protected against sorties by redoubts or trenches near them, and to make this protection easier, or to simplify the artillery service, a very large number of pieces was often massed in one battery. The art of gunnery was still very imperfect. Tartaglia, who was the first to investigate mathematically the motion of

[5] Vere's Commentaries (continuation).

INTRODUCTION. 53

cannon shot, had in his first work (published in 1537) described the trajectory as consisting of three parts: a straight part, due to the impulse of the powder; then a curved part; and lastly, a straight part, due to the weight of the shot. He had afterwards corrected this, and shown that the trajectory was curved throughout, but his earlier notion met with more acceptance, and remained the common theory throughout the next century. St. Rémy, who published "Mémoires d'Artillerie" in 1697, says that (besides ricochet fire, then newly introduced) there are "two different modes of firing guns, namely, at extreme range and point-blank." For the former the piece should be laid at an angle of 45°, and ranges of 4000 yards or more were obtained. But to hit any particular object, he says, it is necessary that that object "should be in the direction or prolongation of the bore of the gun. As the shot, from its weight, continually approaches the ground during every moment of its flight, the distance to be traversed must be so short that it shall not deviate sensibly from the straight line while it is on its way. Experience has shown that this distance cannot be more than 300 toises (600 yards), and this is what one calls point blank range." The use of scales for laying guns at moderate degrees of elevation, though mentioned by a German writer of the sixteenth century, seems to have been practically unknown, at all events in France. To make the most of point-blank range, charges of one-half or two-thirds the shot's weight were employed, and the guns were made long and heavy. The 24-pounder, for instance, was 11 feet long, and weighed 45 cwt.

During the Dutch war of independence, one most important development of artillery took place. Shells and hand-grenades were not indeed invented, but at all events made for the first time practically effective in Holland. At the siege of Groll by Prince Henry of Nassau, in 1627, the shells did a great deal of damage to the houses, and scared the inhabitants; and two years later, at the siege of Bois le Duc, they drove the garrison to abandon two of their outworks, in one case blowing up a powder magazine. Malthus, an Englishman, introduced them into France a few years afterwards.

Hand-grenades had preceded shells, for the earlier attempts to discharge explosive projectiles from cannon had proved discouraging. The fuse being placed next to the charge, to ensure its ignition, the shells had burst in the piece, or at any rate before they reached their object. No satisfactory results were obtained until mortars were introduced, which allowed the fuse to be turned towards the mouth of the piece, and to be lighted separately like a grenade, before the match was applied to the charge. This method of double ignition continued in use down to the middle of the eighteenth century, although several methods of single ignition had been contrived. It was necessary to pack the shell carefully with earth all round the fuse, lest a spark from the fuse or the port-fire should light the charge. The range was varied by varying the elevation, or by increasing the charge when an elevation of 45° was insufficient. A 12-inch mortar would range up to 700 yards with its ordinary charge of 2 lbs., according to St. Rémy's tables, and this might be extended to nearly 1100 yards with a charge of 3 lbs.

As regards the Dutch fortification, its general character was determined by the peculiar conditions of time and place under which it was developed. Called upon to

INTRODUCTION. 55

improvise defences, sometimes, as at Haarlem, with the enemy at the door, the Dutch could not look to masonry for security, or take the citadel of Antwerp for their model. But water was everywhere to their hand and wide and shallow wet ditches, while constituting the best of obstacles, were also the readiest means of obtaining earth for the ramparts. High command was unnecessary for the latter in so flat a country, and even harmful, as it hid the enemy when he reached the edge of the ditch. The more grazing the fire, the more effective was it against every method of crossing. This led them also to make use of the fausse-braye. Screened by the glacis from the besieger's view, it opposed an uninjured front to his galleries when they broke through the counterscarp; and it turned to account the wide berm which was necessary for the stability and for the repair of the main rampart.

The ease with which the water obstacle could be provided led to the multiplication of outworks, and retrenchments, especially as the greater effectiveness of defence at close range, where hand-grenades could be used, made a succession of narrow belts worth more than a wide one. At weak points half-moons, ravelins, *tenailles*, bastioned fronts singly or in pairs (i.e. hornworks and crownworks), were accumulated almost without limit, while the covered way and glacis, with a ditch in front, often formed a continuous outer enceinte. Further, to make up for the absence of walls, and give security against surprise, timber was turned to the utmost account. Palisades, fraises, *chevaux de frise*,[7] and stakes became as prominent in

[7] They got this name, Hexham says, at the siege of Groningen in Friseland, where they served for great use by stopping and hindering the enemy's horse when they came to relieve the town." ("Principles of the Art Military." London, 1639.)

fortress as in camp works. Carpenters were among the most valuable of the reinforcements poured into Ostend, and old ships were used for the retrenchments at Haarlem.

The wet ditch equally influenced the trace of the main rampart. Where a wall was the obstacle, it might be worth while to endeavour to provide a flank for it, which, however short, should be hidden from the enemy, as the Frenchman, Errard, for instance, tried to do. Even in this case the introduction of glancing fire (*tir en bricolle*), by which the besiegers searched out hidden flanks by shots glancing off the escarp of the curtain, made the object almost unattainable. But where a wide belt of water was to be defended exposure was inevitable; and a re-entering front, with long open flanks, allowing of a convergent fire both from flanks and faces upon the head of the advancing causeway, was plainly the most suitable. This allowed also of wide gorges to the bastions, which could be well retrenched.

But slow and costly as it might be, the passage of a wet ditch could hardly be arrested when the water was shallow and stagnant. To produce a current at will, to scour the ditches so as to prevent their silting up, and to vary the water level, sluices were necessary; and Simon Stevin, the teacher of Maurice, was the first engineer to show their value, and to treat fully of their construction.[8]

In France, as elsewhere, in the early part of the seventeenth century, the Italians were gradually making way for native engineers; though the latter could as yet hardly be said to form a school, like those trained under the

---

[8] "Nouvelle Manière de Fortification par Escluses." Leyden, 1618. See also M. Brialmont's remarks in Steichen's Memoir of Stevin.

INTRODUCTION. 57

Nassau princes. There were two in particular—the Chevalier De Ville and the Comte de Pagan—who differed widely in many respects, but were alike in ability and in large experience of sieges. De Ville was not an inventor, but a collector and critic of other men's practice. The fulness and fairness with which every point is treated, the arguments on both sides stated, and instances adduced, make his book (published when he was only thirty-two)[9] perhaps a better guide than any other to the defensive science of his time. The much shorter work of Pagan,[1] —written when he had lost his sight, after twenty years of active service—had for its object to explain and enforce particular proposals by which he hoped to cure the weakness of fortresses. "The strongest do not hold out more than six weeks; the best cannot take care of themselves without an army close at hand; and one no longer asks before attacking whether they are good, but only whether the circumvallation can be finished before the enemy comes up."

His remedy lay in powerful flank defence by artillery, especially for the faces of the bastions. Instead of trying to hide his flanking guns from the enemy's battery on the counterscarp at the salient, he trusted to crushing that battery by a superior number of pieces. His flanks were to be in three tiers, mounting thirteen guns in all; and they were to be perpendicular to the faces they flanked, instead of being perpendicular to the curtain. He preferred wet ditches because they compelled the besieger to make galleries across them under this powerful fire, whereas a dry ditch might be mined under.

[9] "Les Fortifications du Chevalier Antoine de Ville." Paris, 1628.
[1] "Les Fortifications du Comte de Pagan." Paris, 1645.

Pagan was a reformer in advance of his age. The next century adopted the leading features of his trace, but it did not find much favour in his own day. It was objected that the faces of the bastions, which were the parts most easy to attack, were too long; that the size of the bastions made it impossible to fill them with earth, and therefore difficult to retrench them; and that the direction of the flanks exposed them unnecessarily. On these grounds Manesson Mallet,[2] in Vauban's time, recommended a compromise between the trace of Pagan and that of De Ville (which was based upon the Dutch school), and this corresponds with Vauban's own earlier practice.

All the French engineers, unlike the Dutch, clung to orillons, in spite of the length of flank which they absorbed. As De Ville puts it—"the flank is the arm of fortification, and should be clothed and armoured to preserve it for its work." Casemated flanks he altogether condemned, and spoke of them as things of the past, but upper and lower tiers were general.

De Ville's treatise dealt not only with the construction of fortresses, but also with their attack and defence. But the latter subject was more fully handled in a later work of his.[3] It is noticeable that in this work the measures to be taken to guard against treachery and surprise occupy more space than the defence against a regular attack. Surprise was so much dreaded, that where it was attempted and failed he would have no quarter given to the prisoners taken. By employing deception, he argued, they placed themselves outside the laws of war. As regards a regular attack, although the garrison will do well to

[2] "Les Travaux de Mars." 3 vols., Paris, 1685.
[3] "De la Charge des Gouverneurs des Places." Paris, 1639.

annoy the besiegers as much as they can with their artillery, yet all their efforts are not likely to prevent the approaches being carried up to the counterscarp within three or four nights. They should hold the outworks as long as they can; but it is when these are taken and the besiegers prepare to cross the main ditch that the true defence, he says, begins.

Though gunpowder had by this time become effective in all its various applications, for cannon and for small arms, in shells and in mines, real fighting was still at close quarters; and a siege in the middle of the seventeenth century was more nearly akin to a mediæval siege than to a siege of the present day.

But the materials were ready to hand, and the only thing needed was the rise of a man possessing sufficient experience and grasp of mind to combine them into a system, and sufficient authority to bring that system into vogue. The wars of Louis XIV. soon produced such a man and furnished him with unequalled opportunities.

CHAPTER II.

VAUBAN.

The eighteenth century and the first half of the nineteenth were conspicuous for the advance made in the art of war as a whole. Embracing the campaigns of Marlborough, of Frederic, and of Napoleon, it was a period remarkable, not only for the quantity of fighting done, and the quality of the leadership, but even more for the amount of thought devoted to military theory, and for the improvements in strategy, tactics, and organization. This was the case also with some branches of military engineering. Fortification was further developed in the hands of Montalembert, Carnot and the German engineers; while Bélidor furnished besiegers with a new weapon in mine warfare, which largely affected its conditions and its practice. But during all this time, with the exception of underground operations, there was singularly little change in the mode of conducting a siege. It would have seemed preposterous, thirty or forty years ago, to give extracts from the drill-books of Marlborough's day, in order to show how troops should be drawn up; and yet it was almost a matter of course that an article on the attack of fortresses (like that written by the late Sir John Burgoyne for the *Aide-Mémoire* to the military sciences), should " give plans of the regular system of attack as laid down by Vauban, and never altered since, as the best illustration

of the nature of the principal operations." Even now, though breech-loaders and rifled guns have worked changes, the full extent of which no one can foresee, this system in great measure holds its ground. Ricochet fire has lost its value; parallels have changed their distances; it has become necessary for the siege batteries to open at an earlier stage and at undreamt-of ranges; the details of execution of saps and batteries are completely altered; but the general principles which Vauban was the first to grasp, and which his rules embodied, remain as applicable as ever.

Sébastien le Prestre de Vauban was born on the 15th of May, 1633. His father was the second son of the Seigneur de Vauban, and had served in the army, but he was at that time living in very straitened circumstances at the village of St. Léger du Fougeret, near Avallon, in Burgundy. Ten years afterwards he died, leaving his child without home or support; but the boy was adopted and educated by the curé of the village. Before he was eighteen he made his way to Flanders, and enlisted in the regiment of Condé, under a Burgundian captain. Thanks to the curé, he had by that time "a pretty good knowledge of mathematics and fortification, and was not a bad draughtsman," according to his own account; and so before long he was employed as an engineer. For two years he served under Condé, who in the latter part of 1651, in league with Spain, made war against the king. But in 1653 Vauban was taken prisoner, and was persuaded by Mazarin to transfer himself to the royal service. He had attracted notice at the siege of Ste. Menehould, where he swam the Aisne under fire on the day of the assault; and he was now sent to assist in recovering that place for the king.

For his services there he was given a lieutenancy in the Burgundian Foot, but he continued to be employed as before; and in 1655 he received his commission as one of the king's engineers. These did not constitute a corps at that time, but were drawn chiefly from the officers of infantry regiments and retained their regimental commissions. In 1657 he had a company given him in the regiment of La Ferté, and in the following year he was entrusted with the chief direction of the attacks undertaken by Turenne's army, and was warmly commended by Mazarin at the close of the campaign. Eight years of peace followed, during which he was employed upon works at Dunkirk and elsewhere. When war again broke out in 1667 he greatly distinguished himself at the siege of Lille, under the eyes of the king; and he was made governor of the new citadel of Lille, which was built from his designs. During the six years' war which followed the invasion of Holland (1672-8) he had a chief share in seventeen sieges and one defence, and rose to be brigadier and major-general. At its close he was made commissary-general of fortifications, with the chief direction of all works of defence throughout France. In the short war which was ended by the treaty of Ratisbon, in 1684, the siege of Luxemburg gained him fresh reputation.

"I get letters from all sides," he wrote to Louvois, "to congratulate me that the king has had the goodness to make me lieutenant-general: it is even to be seen in print in the Gazettes of Holland, and the historical journal of Woerden; but nevertheless those who ought to know best tell me nothing of it. So, if you please, Monseigneur, let me either be repaid the postage of the eighty or a hundred letters that I have had to pay for, or obtain from his majesty that I should be made lieutenant-general indeed, so as not to give the lie to so many worthy people."[1]

But whether the king objected to have his favours forestalled by public opinion, or hesitated to give such unprecedented promotion to a mere engineer, it was not until four years afterwards that Vauban obtained the rank. The ten years' war which began in 1688, and closed with the peace of Ryswick, took him again into the field and allowed him—especially at Philipsburg, Namur, and Ath, to perfect his method, and surpass his former achievements. In 1702 he had become the senior lieutenant-general, and learning that some new marshals were to be named shortly, he asked the king to include him among the number; or, if the nature of his duties would make that undesirable, at all events to make public the reason for passing him over. This highest dignity was not denied him, and in the beginning of 1703, four years before his death, he was made a marshal of France.

Such were the chief steps upward in his long career. The services by which these steps were earned form so long a list that it would be tedious to attempt to specify them. A year before his death he himself summed them up as follows :—"I am now in the seventy-third year of my age, bearing the load of fifty-two years of service, and the extra load of fifty important sieges, and nearly forty years of incessant journeys to examine fortresses on the frontier which have cost me much suffering and fatigue, both of mind and body, for winter and summer have been alike to me."[2] In forty sieges he had the chief direction of

---

[1] Michel, "Histoire de Vauban." Paris, 1879, p. 196.
[2] Id. p. 359.

the attacks, and in every one of these he was successful. He was twice engaged in the defence of fortresses; at Condé in 1656, and at Oudenarde in 1674. In the former case the garrison had to surrender from want of provisions; in the latter the siege was soon raised. He is said to have designed or amended the works of more than 160 fortresses, among which may be mentioned Dunkirk, Menin, Landau, Neuf-Brisach, and the citadels of Lille and Strasbourg.

It is not surprising that a man who brought to this extraordinary range of experience a remarkable capacity for turning experience to account, a singularly cool and sound judgment, and a freshness of mind that was proof against age, should have attained an unique position among military engineers. The chances of war, at no time very favourable to engineers, were in those days especially adverse. Vauban himself styled them "the martyrs of the infantry." In his later years he wrote:—

"Formerly, men of that profession were very scarce in France, and the few there were lasted so short a time that it was still more rare to meet with any who had seen five or six sieges, and rarer still to find any who had done so without receiving several wounds, which disabling them at the beginning or in the middle of a siege, prevented their seeing the end of it, and so gaining skill." [3]

Besides, the direction of attacks formed only one branch of the engineer's duties. As Vauban himself said in a letter in 1693:—

"I could teach any officer of common sense to manage an approach, a lodgment on the counterscarp, a descent

[3] "Traité de l'Attaque des Places."

into the ditch, an attachment of the miner, &c., in the course of three average sieges; but a good builder is only to be made by fifteen or twenty years of application, and even then he must have had a variety of employment, and be a very hardworking man. We have at present a good number of men who are fit for sieges, but very few who thoroughly understand building, and still fewer who understand both one and the other.... Engineering is a business beyond our strength; it embraces too many things for a man to be able to make himself perfectly master of it: I think so well of myself as to believe that I am one of the strongest of the lot, and capable of giving lessons to the most skilful of them, and yet with all that, when I examine myself, I find myself not more than half an engineer, after forty years of very hard study, and of the largest experience any one ever had. Thanks be to Him who has preserved me, and let me live till now!"[4]

He was himself repeatedly wounded—five times in his first five years of service, and three times afterwards, notwithstanding the solicitude of which he was latterly the object. In 1677 Louvois wrote to Marshal d'Humières who was about to besiege St. Ghislain:—"His Majesty is willing that you should take M. de Vauban with you, but strongly urges upon you to take care of him, and not to allow him to assume the direct conduct of the approaches." And in 1683 Marshal d'Humières, having again obtained Vauban's assistance for the siege of Courtrai, wrote apologetically to Louvois:—

"I have not been able to prevent M. de Vauban from going into the town" (during the attack on the citadel;)

[4] Augoyat, "Aperçu sur les Ingénieurs, i. 216."

"he promised me faithfully that he would not stir out of his lodging, but would receive reports there from his engineers of what was going on. I even charged the Marquis d'Huxelles not to leave him, and to prevent his going near the citadel. We have been afraid that we should get into trouble about this; but you know that one cannot manage him just as one pleases, and if any one deserves to be scolded, 'I assure you it is not I." [5]

Incapable of courting danger for the mere display of courage, he was apt to expose himself in his anxiety to observe the enemy's works. The best way of reconnoitring a place, he says, in order not to attract attention and draw fire, is to have one's escort concealed at a little distance behind, and to go forward alone, or almost alone; "that is what I have almost always done, and I have found it succeed." At Luxemburg he advanced in this way night after night up to the palisades of the covered way. Once he was discovered, but he made a gesture to the besieged not to fire, and walked onward instead of retiring. They concluded that he must be one of their own men, and allowed him to finish his observations, and to make his way back untouched.

The minister's anxious concern for his safety went along with unceasing demands on his exertions. During peace he was perpetually travelling from one part of France to another, inspecting works in progress or designing new ones. For instance, in 1681, after visiting Besançon, Phalsbourg, and Schlettstadt, in the east of France, he was by midsummer in the Isle of Rhé, on the west coast, planning a citadel and enceinte. After paying

---

[5] Michel, p. 184.

a visit to the harbour works at Toulon, he reached Strasbourg in October, at the moment of its seizure by Louis XIV. By the middle of November he had prepared his project, consisting of a large volume of manuscript and seventeen sheets of drawings, and providing for a new citadel and various improvements. He was hurried away from there by Louvois to Casale in Piedmont, where he arrived in the beginning of 1682.

He was married in 1660, but for the next fifteen years his wife lived with her parents. In 1675, having obtained a short leave of absence for the first time for nearly ten years, he purchased the estate of Bazoches, near Avallon, and built a château there. This was henceforth his home, but his visits to it were brief and rare. It was not till after the peace of Ryswick that he enjoyed any real leisure.

Of his labours at sieges he has left some vivid pictures. After the fall of Luxemburg he writes to Louvois :—" If you do not give me two or three days' rest after the siege, I am done for; at this moment I am so weary and so sleepy that I don't know what I say." But almost immediately afterwards he was on the road for Versailles, to receive instructions about the creation of the park and gardens, upon which the troops who had taken Luxemburg were set to work, and where they lost in a few months more men than they had done in the siege. Again, during the siege of Philipsburg in 1688, Vauban excuses himself to Louvois for not writing more frequently.

"I am overwhelmed," he says, "with work, and it is not possible to visit daily two attacks, where one has to look and look again into I don't know how many different things, to argue, to detail, to give the same orders ten times over, and to spend an hour and a half or two hours every day in reporting everything to Monseigneur, to write to this man and to that, and a thousand other details that one has to go into, which make the days always too short to my mind; though my body on the other hand, finds them full long. For if all our trenches were put end to end, they would form a straight line of six good miles, of which I traverse every day more than two-thirds, usually with wet feet, and over a hundred thousand fascines, which have been used to pave the trenches, and which are about as easy as logs to walk over: judge of the pleasantness of the promenade."

Yet the day after Philipsburg surrendered, Vauban was already on his way to Manheim, which was to be next besieged.

Neither rank nor age quenched his activity. When a marshal, he consented to serve as chief engineer at the siege of Brisach; and though seventy years old and suffering from chronic bronchitis, he writes:—

"I am well enough satisfied with my last night, which I partly spent in searching the bends of the Upper Rhine, which may help the attacks on that side. I have found some very favourable sites there for reverse and ricochet batteries, which, please God, I shall take advantage of during the siege. It was daytime before I came away well soaked by a light mist." [6]

Even slights and mortifications could not check his eagerness to be serviceable. After the fall of Brisach it was determined to lay siege to Landau, and Vauban, on hearing of it, wrote to the minister:—

[6] Michel, p. 348.

"Old as I am, I do not yet sentence myself to repose, and when it is a question of rendering an important service to the King, I shall be ready enough to put all considerations on one side, whether as regards myself or as regards the dignity with which he has been pleased to honour me, persuaded as I am that anything, however small, is honourable if it goes to serve the King and State; much more when it admits of such considerable service as I could render in the siege in question. On this account, although it is little to be desired for myself, since apparently it will be cold, wet, and lengthy, and there are many murmurs about the discomforts of the season we are entering on, and the postponement of winter-quarters, of which the troops have so much need, I pass lightly over all these considerations, just as over that of my own dignity, and I offer with all my heart al my practical experience to the King, in whatever capacity he may think fit. If I can succeed in satisfying him, I am sure to be satisfied myself. Therefore, sir, let me know his will; the sooner the better, for it is of no use to offer oneself, and even to throw oneself at him, if one is not accepted. What compels me to speak to you in this way is that there seems to me to be an intention of conducting the siege without me. I confess that I am hurt at this."[7]

The fact was that Marshal Tallard, who was to command at the siege, wished to have the credit of it himself and feared that the presence of the great engineer would deprive him of it. But when Vauban learnt that he was not to be employed in the recovery of this place, which was one of his own masterpieces, he drew up an elaborate

[7] Michel, p. 352.

memoir on the best mode of attacking it. "I wish to console myself as best I can," he writes, " by imparting my ideas and knowledge to those who are to take my place, in order that I may at least have the satisfaction of not being altogether useless to his Majesty in an affair so important as this seems likely to be." During the siege of Turin, in 1706, he showed himself equally ready to labour for the success of an enterprise which he was not able to take part in; but his advice was disregarded by the presumptuous La Feuillade, and the siege ended in failure and disaster. Yet he was far from being insensible to slights. Shortly before the siege of Valenciennes, in 1677, he wrote to Louvois :—

"It is rather a curious thing to see that every one knows what you intend to do, and that it is only to me that any secret is made of it; apparently I am to play an insignificant part in it, and my opinion is to count for nothing. Thank God, I will do my duty; but I will take care not to undertake all I have done at other sieges. I promise you that." [8]

Though sharing most devoutly in the monarch-worship of his age and country, he was honourably distinguished by his self-respect and independence of tone. Louvois—a firm friend to him, but passionate and overbearing—repeatedly urged him, when he was fortifying Dunkirk, to substitute a redoubt for a certain hornwork. At length Vauban remonstrates:—

"Settle what you please on that point by way of authority, but don't attempt to convince me by reason, since I have that altogether on my side ; and in God's name let us have done with quibbling, for henceforward

[8] Michel, p. 113.

I will not spend another word in argument about the redoubt or the hornwork." [9]

In 1671 it was alleged that frauds had been practised on working parties of the troops by engineers under his orders, and Louvois called for a report from him. Vauban warmly vindicated his subordinates, and took the chief responsibility on himself. He, if any one, deserves to be punished; or if he is innocent, then his accusers.

"And as to that, Monseigneur," he adds, "I will take the liberty of telling you that affairs have gone too far to stop here; for I am accused by persons whose names I shall find out, who have spread villainous reports about me, so that it is necessary that I should be most completely justified. In one word, you quite understand that unless you should go to the bottom of this affair you could not do me justice, and in failing to do me justice, you would oblige me to look out for means of doing myself justice, and to abandon for ever fortification and all its connections. So make a bold and strict examination, without any partiality; for I tell you freely that, relying upon a scrupulous honesty, I fear neither the King, nor you, nor all mankind put together. Fortune has made me by birth the poorest man of quality in France, but, to make up, it has favoured me with an honest heart, so exempt from every kind of rascality, that it cannot bear the mere thought of it without horror." [1]

St. Simon, who was no panegyrist, has described him as "perhaps the most honest and most virtuous man of his age; and with the reputation of being the most skilful in the art of sieges and of fortification; the most simple-

---

[9] Michel, p. 256.   [1] Id. p. 262.

minded, most truthful, and most modest." He was of middle height, strongly built, and of a hardy constitution, of a rough and soldierly bearing which seemed to denote a harsh and inflexible character. "But nothing could be further from him," says St. Simon; "never was man more gentle, more kindly, or more obliging."

It was the experience and authority which Vauban acquired by his extended service that alone enabled him to mature his improvements in siege operations. If, like his distinguished predecessor Pagan, he had been incapacitated for active life before he reached the age of forty, his name would hardly have stood so high as Pagan's. He had served twenty-two years when he first made use of parallels, thirty-seven years when he first tried ricochet fire, and forty-six years before he was able to exhibit them both in full efficiency and in combination. But the general principles of the method of which they formed the most striking features had taken root in his mind much earlier.

In 1669 Louvois, annoyed at his own ignorance of an art, with which as War Minister he had so much to do, called on Vauban to give him some account of it. The "Mémoire pour servir d'instruction dans la Conduite des Siéges,"[2] which Vauban composed for him, hurriedly written as it was in the short space of six weeks, is of the highest interest; both as a picture of the siege-warfare of that day, and as the starting-point of his own reforms. "Nothing," as he afterwards told Pellisson, "had ever been so useful to him as this attention and close consideration, pen in hand, of all that he had ever thought of or seen on

[2] It was published at Leyden in 1740, and was erroneously described on the title-page as the treatise presented to Louis XIV. in 1704.

this subject," and it was at this time that he shaped and settled the method of attack which he afterwards put in practice.[3]

In this memoir Vauban begins by enumerating the mistakes then commonly made in sieges. Among these he dwells particularly upon the confused and unsystematic character of the attacks.

"Men work on from day to day without ever knowing what they will do two hours hence. So that everything is done in a disorderly, tentative way; from which it follows that an approach is always ill-directed. The batteries and places of arms are never where they ought to be; proper arrangements are never made for establishing the lodgments; one is never in a favourable position for meeting a sortie; and it never, or hardly ever, happens but that the approaches are longer by one-half or one-third than they need have been, and that after all they are enfiladed somewhere."

All this is mainly due to the interference of the general commanding in the trenches for the day.

"The emulation between the general officers often leads them to expose their soldiers to no purpose, trying to make them do more than they can, and caring little if they get a score or two killed so long as they can obtain four paces more progress than their fellows. By their authority they direct the course of the approaches as they please, and are continually interrupting the plan of attack and all the arrangements of the engineer, who far from being able to follow the systematic action which would have brought affairs to a good end, finds himself reduced to serve as an instrument of their varying

[3] Pellisson, " Lettres historiques," iii. 270.

caprices. Varying I say, for one commands one way to-day, and to-morrow the general who relieves him will command quite another way; and as they are not always endowed with the highest capacity for matters of this sort, God knows what failures and what waste they cause, and how much needless blood they shed in the course of a siege. But what is most absurd is to see these gentlemen, when they have been relieved in the trenches, describe and lament, or rather boast with a self-satisfied and complacent air, how they have lost a hundred or a hundred and fifty men during their turn of duty, among which perhaps there will be eight or ten officers and some brave engineers, who might have done service elsewhere. Is not that something to be pleased at? and is not their prince much indebted to those who obtain with the loss of a hundred men what might have been obtained by a little industry with a loss of ten? In truth, if states perish for want of good men to defend them, I know of no punishment severe enough for those who rob them of them to no purpose."

This plain and strong language was not thrown away. At the siege of Maestricht, four years afterwards, the king entrusted to Vauban the sole direction of the siege-works, and restricted the functions of the generals of the day to the command of the guard of the trenches.

But the fault did not lie wholly with the generals, as Vauban went on to explain in his memoir:—

"To change the present system in the trenches, there is need of new instructions; need of engineers who have a strong hold of firmly established principles; of workmen specially trained and taught; of materials sufficient in quantity and good in quality; and above all of a fixed and constant resolution not to depart from rules which have been once laid down, when their soundness and utility have been verified by reason and experience."

To illustrate his criticisms on the mode of conducting sieges at that time, he took as an example the attack which he had himself directed against Lille two years before; which, as he says, " met with much approbation, and in which, to tell the truth, there were fewer inutilities than in any other for a long time past." He points out faults committed at every stage of this attack, and contrasts it with an imaginary attack upon the same front in which these faults are corrected.[4] He particularly blames the want of proper support for the saps, and the position of the batteries (for which he was not himself responsible), at too great a distance from the fortress, and not far enough apart to give any real convergence of fire. And yet this was reckoned the best-managed attack in a siege where the king was present in person, and where there was no lack of men, money, munitions, or good engineers. It must be regarded as a very favourable specimen of the usual procedure. Hence he concludes that " when we succeed, it is rather owing to the weakness of the enemy than to our own merit. I leave others, then, to judge whether it is important to remedy these defects, and to seek means of reducing the conduct of sieges to a more systematic and a less bloody method."

Such a method he goes on to describe in detail; a method by adopting which he is bold to affirm, that " one would save more than three-fourths of the men one usually loses, one would avoid much useless expense, one would

[4] Vide Occasional Papers of the R.E. Institute. Vol. vi. Paper No. 2, Plate 4.

be always safe, one would get on quite as fast as if one hurried, and lastly one would be certain to succeed in undertakings which now in most cases prove failures." He gives rules for placing the batteries, and for every stage of a siege. He connects the approaches by two extended "places of arms, at some distance from the fortress;" but the point on which he lays most stress is, that another place of arms should be made near the foot of the glacis, six yards wide and 600 yards long, overlapping the heads of the approaches. The time spent in the formation of this third parallel (to call it by its later name), will be more than regained by the help it will afford in making the lodgments upon the crest of the glacis, apart from the saving of life. For "it is well established that in the sieges of places that make a good defence one loses three times as many men before the capture of the counterscarp, as one loses from that time up to the surrender of the place. This loss is always due to over-eagerness; we do not take half the precautions that such an enterprise requires, and, as a necessary consequence, instead of gaining one day we lose two, at the expense of our best soldiers who perish miserably on such occasions."

For infantry and artillery alike it is his constant aim to secure an enveloping position. "The attack which is able completely to envelop the front of a place attacked is preferable to all others. And on the contrary the worst of all attacks is that of which the head is enveloped by the front attacked." These axioms of the memoir form the basis, not only of the rules which follow them, but of the modern art of sieges.

His first opportunity of carrying out his principles in at all a complete form was at Maestricht in 1673. There, as

Louis XIV. himself describes it in his memoirs:—"We went towards the place as it were in order of battle, with grand parallel lines, wide and spacious; so that, by means of the steps in them, one could march upon the enemy with a broad front."[5] The Governor of Maestricht said that "it had fallen to him to stand six considerable sieges, but that he had seen none like this; and that from the first day he had lost hope of being able to do anything, seeing how the guards of the trenches were supported, and that there was no means of making a sortie without being cut to pieces; that the man who directed the approaches must be the most skilful man in the world."[6]

Vauban was assisted at Maestricht by an engineer named Paul, who had taken part in the long defence of Candia, which ended in 1668; and it has often been asserted that it was from the Turkish siege-works before Candia that the parallels of Maestricht were borrowed. According to Pellisson, indeed, Vauban himself admitted this. Pellisson was a hanger-on of the Court, and while attending Louis XIV. in his campaigns, kept a diary in the form of letters. He wrote just after the siege that the attack "has something of the air of those made by the Turks before Candia, and one can trace some sort of imitation of their method;" and four years afterwards, speaking of a conversation he had had with Vauban at Tournay, he says:—"He owned to me that he had changed his mode of attacking at the siege of Maestricht, and in fact that he had copied the Turks and their works before Candia, with their numerous lines parallel to the place, which is what I had myself remarked some time ago."[7]

[5] Allent, "Histoire du Corps du Génie," p. 108.
[6] Pellisson, "Lettres historiques," i. 362.
[7] Id. iii. 370.

But he adds that Vauban went on to say that the change was due to the memoir on sieges which he had had to write for Louvois in 1669, for that by his reflections upon the subject then, "he settled the method of attack which he now carries out."

It seems not improbable, therefore, that Pellisson, in his eagerness to get a confirmation from Vauban of his own original surmise, may have somewhat overstated the case; and that the innumerable and unsystematic parallels of the Turks had little to do with the evolution of Vauban's method, though they may have helped to gain it acceptance by giving it something of a foreign flavour. Vauban makes no reference to them in his memoir, and his own personal experience seems to have been the basis of his reforms.

The idea of occasional parallels was, in fact, already afloat in France, although a master-hand was needed to develop its value. The plan of the siege of La Capelle (in 1637) shows "a grand place of arms parallel to the front of attack, embracing the two collateral half-fronts, its right resting on a redoubt, and with the batteries disposed upon it."[8] D'Aurignac, in a work published in 1668,[9] gives a scheme of attack which he declares that he put in practice with much success when directing the attack on Bellegarde. Just beyond musket-range of the outworks of the place there is a grand place of arms capable of containing two battalions and a squadron. From its extremities the approaches are made; and at every fifty yards of advance these are crossed by other places of arms,

---

[8] Augoyat, "Aperçu sur les Ingénieurs," i. 55.
[9] "Livre de toutes sortes de Fortifications." See also Occasional Papers of the R.E. Institute. Vol. vi. Paper No. 2. Plate 3.

fifty yards long. As each successive place of arms is finished, half a battalion is to be moved up into it as a guard for the workmen. The guards, he says, should always be "kept in a body in the places of arms, and not broken up in the approaches, as it has been the custom to do," in which case they are sure to be routed by sorties. Lastly, at about twenty paces from the salient of the counterscarp of the ravelin, the approaches are once more connected by "trenches parallel to the place," the support of which will allow lodgments to be made simultaneously upon the counterscarp of the ravelin and of the bastions on each side of it.

It seems plain, then, that the importance of presenting a broad front to check sorties was already beginning to be recognized by engineers. Vauban's merit lay not so much in the idea itself, as in the boldness and judgment with which he applied it. He took care that the approaches should be so "escorted by places of arms" (to use his own expression) that attacks upon them could be quickly repulsed; and he gave these places of arms an extension which not only furthered their own function, but also gave opportunity for dispersing the batteries and converging their fire. At the same time he avoided any excessive frequency, or slowness in execution of them, which might have brought them into discredit; and he contrived to astonish every one by the rapidity of his advance.

At Maestricht, which was taken within a fortnight, Vauban made three parallels, having a length of from 600 to 800 yards each. In subsequent sieges there was, he says, little uniformity of practice as regards them, owing to the want of definite rules; and they were sometimes badly placed. But whatever irregularities accident or special circumstances occasioned in the trace of his attacks, almost every siege furnished new illustrations of his principles, and especially of his leading principle—to rely on art and industry rather than on force. At Cambrai, in 1677, Louis XIV. insisted on assaulting a ravelin against his advice. "Sire," he said, "you will lose lives there that are worth more than the ravelin." The troops carried the work, but were driven out again with loss; and Vauban was then allowed to push on his approaches and take it in his own fashion, which he did with a loss of five men. "I will believe you another time," the king said; and he allowed Vauban to dissuade him from his angry purpose of refusing terms to the garrison.

Yet before this time, at Valenciennes, Vauban had shown that he could on occasion be bold beyond others. As the glacis was countermined, he proposed, contrary to his usual custom, to carry the covered way by assault; and he recommended that this assault, in which several thousand men would be engaged, should be delivered, not at night, as was customary, but in broad daylight, when the enemy would be less on the alert, and there would be less risk of confusion and misbehaviour on the part of the troops. The most experienced generals, Schomberg, Luxemburg and others, were against this proposal, but after much argument the king consented. The assault met with unhoped-for success, and not only the covered way, but the place itself was gained, with a loss of less than fifty men.

But this success did not tempt him to employ assaults where other means were open to him. At Luxemburg, in 1684, the covered way, provided with masonry keeps threatened to prove more troublesome than usual. Instead of using force, he stopped the sap just out of range of hand-grenades, and built up parapets ten feet or more in height with successive tiers of gabions. From these *trench cavaliers*, here used for the first time, he was able to plunge into, and enfilade the covered way, and to dislodge the enemy from the more advanced parts of it.

And as with the covered way, so with the breaches: he always preferred, if possible, to gain possession of them step by step. At the siege of Charleroi, in 1693, he was at first blamed for having chosen what seemed to be the strongest side as the point of attack. But before long his choice was vindicated, the outworks were taken, the body of the place breached, and the troops became impatient for the assault. Yet though the murmurs of the camp at his over-caution were echoed back from the court, Vauban was obstinate. "Let us burn more powder, and shed less blood," he replied, and continued at work with his miners until the garrison, who had themselves mined the bastion in readiness for an assault, found further resistance hopeless, and surrendered.

In 1692, at the siege of Namur, Vauban found himself face to face with his rival Coehorn, who was defending a fort of his own construction, but was obliged to surrender. Three years afterwards Coehorn himself directed the attacks, when William III. recovered the place. With a strong likeness in their general course, the two sieges presented some marked contrasts, very characteristic of the two engineers:—

"Vauban, employing no more guns than were necessary, using all his influence to restrain the troops, not allowing them to advance except under cover, and bringing them in this way to the foot of each work, had made it his study and his pride to spare them; and had done this without slackening the siege: Coehorn accumulating ordnance, sending the troops across the open to make assaults at a distance, and sacrificing everything to his eagerness to shorten the siege, and to scare and frighten the defenders, had economized neither money, nor men, nor in fact time."[1]

The siege of town and castle had occupied five weeks in 1692; it occupied two months in 1695. The loss of the besiegers, which was under 3000 in the former case, was nearly 9000 in the latter. At the same time, allowance must be made for the fact that in the second siege the place itself was stronger, and the garrison larger. When Vauban heard of the general assault in the second siege, in which the English Grenadiers crossed nearly half a mile of open ground with drums beating and colours flying, on their way to the breach, he wrote: "I never saw anything like it, or even approaching it; for the magnitude of the blunder, I mean, not for the grandeur of the action, for I find that too senseless to admire it."

He had the same aversion to random violence in the artillery, as in the engineer, operations of a siege. Bombardments, which were much to the taste of the harsh and impatient Louvois, met with uniform opposition from Vauban. "Never fire at the buildings of fortresses," is one of his maxims, "for that is to lose time and waste ammunition for things which contribute nothing to their surrender, and the repairs of which always cost you much after the place is taken." And elsewhere he says:[2] "One

[1] Allent, "Histoire du Corps du Genie," p. 317.
[2] "Traité de l'Attaque," pp. 263 and 122.

should fire merely at the works and batteries of the fortress, and into the centres of the bastions and ravelins where retrenchments may be made." Even for this use of shells he was not lavish of them, though he thought highly of their effect. In his estimate of ammunition for a siege the shells were only about one-fifth of the shot; and he speaks slightingly of the small mortars for deluging the works with grenades which had been introduced by Coehorn, and of which Coehorn employed no less than 500 at the siege of Bonn. He protested against the waste of ammunition by opening fire at long ranges. In his early memoir of 1669 he said that the main gun-batteries should seldom be more than 400 or less than 300 yards from the counterscarp: "at this distance the shot has almost its full force, and if one were to bring them nearer, their construction would be too long delayed." But he had noticed, and he pointed out in this memoir, that "enfilade fire from a distance is more annoying than from close at hand, because the violence of the shots which come from a distance being abated and almost exhausted, the balls drop away from the straight line; whence it follows that the traverses one provides against them, however high, cannot prevent their plunging between them. When on the contrary the fire comes from close at hand, it is not very difficult to protect oneself from it, since the shortness of the range causes the ball to be impelled with such violence, that it deviates little or nothing from the direct line; whence it further follows that if it grazes the top of one traverse, it will be stopped by the mass of the next, without doing any damage between them."

These remarks had immediate reference to the approaches of the besieger, but they applied equally to the works of the besieged, and Vauban set himself to combine the advantages of short range and of highly curved fire by using reduced charges and giving increased elevation.

It was at Philipsburg, in 1688, that he made his first essay with ricochet batteries. Of one of these he told Louvois that it "dismounted six or seven guns, and caused one of the long sides of the hornwork and the whole of the face of one of the bastions opposite to the main attack to be so deserted that their fire quite ceased." A few weeks later he again wrote to Louvois, after the capture of Manheim, that his ricochet battery there "had only fired one day and had dismounted four or five pieces of artillery, made the defenders abandon six or seven others, set fire to five or six shells, and to two casks of powder which made the hats fly up in the air, taken off the leg of a lieutenant-colonel, and persecuted I don't know how many people whom it hunted out of nooks where nothing but the sky was to be seen."

But it was at Ath, in 1697, that he gave his grand demonstration of the effect of this and of all his other improvements in the art of attack. Ath was a strong place, having been fortified by Vauban himself, but the defence was passive; and the siege-works went on, we are told, "with so much method on our side, and with so little interruption on the side of the enemy, that it seemed rather the representation of a siege, than a siege itself."[3] Vauban himself wrote, "I do not believe there was ever a regular siege, such as this, in which so excellent a place as that which we have just

---

[3] "Journal of the Siege of Ath," attached to "Goulon's Memoirs," and probably written by Vauban's nephew. (Translated by J. Heath. London, 1745.)

taken has been reduced so quickly and with so little loss." It occupied only 14 days, and cost the besiegers only 300 men, killed and wounded.

In the project for the attack it is laid down :—" The first parallel will be called *contravallation*, for its action is the same as that sort of line, but in a manner more sure and more close. It receives all the guard of the trenches. The second parallel will be called *line of the batteries*, for it is on it that we place all the first batteries that are made to subdue the fire of the defence. It supports the saps and trenches. When made, it receives two-thirds of the guard of the trenches. The other third remains in the first, on the wings and in the middle." The first parallel was about a mile and a quarter in length and something under 600 yards from the counterscarp. The second parallel was nearly as long, with its extremities resting upon the first parallel, but not more than 300 yards, in the middle, from the salients of the ravelin and bastions attacked.

" The batteries were placed in a quite different manner from all before them ; for, taking in the whole front of the attack, they traversed and enfiladed with plunging fire the bastions, ravelins, and covered ways of the place in such a manner, that after they were once well in play, the enemy could no longer stand to their defences ; and they so effectually extinguished the fire of the place that the besiegers could pass and repass between the camp and trenches without danger. It was not without difficulty that M. de Vauban prevailed on the officers of the train to lower the charges of their great guns, to batter *à ricochet* with small charges, the effects of which did not presently appear to them ; but after a good deal of painstaking, they were at last reconciled to it. Bounce and clatter and readiness for action had hitherto composed the whole merit of the train at sieges; here the thing was altered, for never was known before so little noise made with so considerable a number of cannon as were fired at this siege. . . . We found after the place was taken that the greatest part of the wounded had their legs and arms carried away upon the rampart by the effects of these batteries, the bullets giving the enemy incessant disquiet on all sides, following them even into their safe retreats, dismounting their guns by breaking the wheels and cheeks of the carriages." [4]

Although in these sieges ricochet-fire proved very effective against guns, this was not in Vauban's eyes the work for which it was most appropriate. "So long as the object is to dismount the enemy's artillery one may fire with full charges," he says; "but as soon as it is dismounted ricochet-fire must be used." He explains that the work of the latter is to drive the enemy's troops from the faces or flanks which might oppose the besieger, to sweep the ditches and communications, to clear the covered way and splinter its palisades; and that it will do this work more certainly, more quickly, and with much less expenditure of powder than any other kind of fire.

It went in fact hand in hand with parallels to secure the besieger's workmen against sorties.

"It is certain," he writes, "that if one establishes places of arms, as proposed in these memoirs, and the troops are properly disposed in them, the enemy will not be able to make sorties without coming in collision with the whole guard of the trenches; and that if on the

[4] "Journal of the Siege."

other hand the ricochet batteries are well served, he cannot assemble troops in any part of the covered ways opposite to the attacks. *Hence few or no sorties.*" [5]

In 1703, when he was seventy years old, and had just returned from the last siege in which he was engaged, he wrote the "Traité de l'Attaque des Places" which has been already referred to, not intending it for publication, but for the use of the Duke of Burgundy, the grandson of Louis XIV. "May it please you," he says, "to keep it for yourself, and to let no one else have it, lest copies should be taken of it which, if they chanced to pass into our enemies' hands, would perhaps be welcomed more than they deserve." Nearly three years afterwards he supplemented it by a "Traité de la Défense des Places," written only a few months before his death. In these two works we have a digest of his vast experience, and his principles, disengaged from particular applications, are presented in their most mature form. The latter are summed up in thirty maxims of which the general substance is as follows:—

To be well informed of the strength of the garrison; to be careful to attack upon the weakest side; not to open the trenches till everything is ready; to embrace the whole front of the works attacked; never to attack reentering angles where the besieger may be enveloped, instead of enveloping; to employ the sap directly open trenchwork becomes dangerous, and "never to do uncovered and by force what can be done by industry, since industry is always sure, whereas force is apt sometimes to fail, and usually runs great risks;" not to push forward the trenches until those that are to support them are

[5] "Traité de l'Attaque," p. 11.

ready; to provide three grand lines, or places of arms, of due extent; always, if possible, to enfilade the works attacked or take them in reverse, by ricochet fire, and gain possession of them by this means instead of by assaults in force; to avoid all precipitation, for that "does not hasten the taking of places but often retards it, and always renders the scene bloody;" not to waste ammunition in bombarding the town; to deviate from regularity in the attack no more than is strictly necessary, and never on the ground that the place is not strong; and to let the chief direction of all the operations, both artillery and engineer, be in one man's hands, under the authority of the general commanding.

Vauban's improvements in the mode of attacking fortresses were the most considerable and the most lasting of his services to the art of war, and he put a seal to them by writing his treatises. In fortress-building his labours were immense, and his work of the highest value, both on account of his skill in dealing with local conditions, and of the order and economy which he introduced generally. But he did not leave his mark on the art of fortification in a corresponding degree, nor has he put on paper with the same completeness his ideas respecting it. "The art of fortifying," he said, "does not consist in rules and systems, but solely in good sense and experience;" and when he was urged to write something on the subject, he answered: "Would you have me teach that a curtain is between two bastions, that a bastion is composed of an angle and two faces, &c.? that is not my line."

But in submitting his project for Landau in 1687 he wrote: "I have taken the opportunity of this project to propose a *system*, which though it has some appearance of novelty is really only an improvement of the old." This was the tower-bastion system, which he afterwards employed in an improved form at Neuf-Brisach. It may be said to be a combination of the bastioned trace with the polygonal or caponier trace, which has since so largely superseded it; the latter being used for the body of the place, and the former furnishing an envelope by which the towers or caponiers are shielded. Vauban himself allowed that "it is quite new and has not yet reached all the perfection that is requisite;" but it would be generally admitted now that his successors would have done better to improve its details, than to turn their backs upon it altogether, and treat it as a whim of his old age. The siege of Landau by the French in 1703, and their defence of it in the following year, said much for the merit of the system. In the former, at the end of a month the besiegers had established themselves on the detached bastions or counterguards, but Marshal Tallard thought it best to offer favourable terms to the garrison lest they should continue to hold out. In 1704 the French held out for seventy days, and similarly capitulated when the besiegers were in possession of the detached bastions.

But with Vauban fortifying meant something more than fortress-building. He may be said to have been the first engineer who considered fortresses collectively, as units in a general scheme of frontier defence. In 1678, after the peace of Nimeguen, he drew up a memoir on the defence of the north-east frontier, in which he recommended the construction of a few new places, in order to provide a double line of fortresses between the Meuse and the sea—a distance of 120 miles, with thirteen places in each line. In carrying out this scheme he took care to secure command of all the roads and watercourses perpendicular to the frontier; and behind his fortresses he carried roads and canals parallel to the frontier, serving as lines of communication for the defence, or of obstacle for the enemy.

He recommended that many of the old fortresses in the interior should be dismantled or demolished; but in a memoir, written probably in 1689, "on the importance of Paris to France and the means that should be taken to secure her," he strongly urged the fortification of the capital. He proposed, while restoring the old enceinte of the city, to construct a new enceinte about a mile or a mile and a half in front of it, occupying the heights of Belleville, Montmartre, &c., as was actually done a century and a half later; and since "a town of this size so fortified might become formidable even to its master," he further proposed to make citadels on the banks of the Seine.

Another favourite idea, which he repeatedly insisted on, and which has since received a great development, was to attach intrenched camps to fortresses, in order that the investments of the latter might be hindered and their defence prolonged. He made a camp of this kind at Dunkirk in 1693 for 11,000 men, with a continuous line of field profile about five miles long. He proposed a similar camp for Namur in 1695, and again for Thionville in 1705. "I know," he wrote regarding the latter, "that this is not to the taste of the King or of his generals, who have given him an unfavourable impression of intrenched camps; but that is because they do not understand them." In order to enlighten them, he had asked a Flemish gentleman in 1693 to hunt up examples of the successful use of field fortification by the Hussites and Turks, and in the Thirty Years' War. "For although I know very well the value of intrenched camps," he says, "I stand in need of the authority of all the great men to recommend them to our foolish nation, which thinks that one ought always to fight just as one is, without any other concern than to hit hard." In 1705 he wrote part of a treatise on field fortification, in which he brought forward these instances; but it seems to have been coldly received, and was never finished.

If he had lost every other title to fame, the name of Vauban would still deserve to be remembered as the inventor of the socket-bayonet. In the latter half of the seventeenth century the flint-lock fusil was gradually displacing the more cumbrous musket. In France, the war ministry opposed the change, and proposals for the improvement of the service weapon which were twice made by Vauban himself were not adopted. But the troops showed their own opinion unmistakably, and after the battle of Steinkirk (1692) the French musketeers and pikemen alike threw away their own arms to take instead the fusils of their beaten enemies. As the proportion of pikes to muskets became less and less, bayonets with wooden hafts to fit into the barrels were given to the musketeers, to enable them to defend themselves in hand-to-hand fighting; but these made the weapon useless for the time as a fire-arm. In 1687 Louvois wrote to Vauban :—

"I have seen officers who have made the campaign in Hungary this year, and who have assured me that in the infantry of the Emperor there is not a single pike; that each battalion is of 400 or 500 men, and the soldiers carry with them chevaux-de-frise, which they connect together and place along the front of the battalion when they are in presence of the enemy." [6]

He asked Vauban to have some such chevaux-de-frise made, and to give him his opinion about them. A fortnight later he wrote again, and his letter indicates what Vauban's reply had been :—

"The King will be glad that when you come here you should bring with you the soldier's equipment you speak of in your letter. But I beg you to explain to me how you contrive a bayonet at the end of a musket which does not prevent one from firing and loading, and what dimensions you propose to give to the said bayonet."

It was at Vauban's instance that at length, in 1703, Louis XIV. decided to abandon the pike altogether, and adopt the fusil and bayonet as the weapons for the whole of the infantry.

He drew up an admirable project for the reorganization of the artillery service, but it was not carried into effect till long after his death. With somewhat better success he urged the formation of standing companies of sappers and miners, to be permanently attached to the engineers. He was unsparing in his efforts to improve the position, and to heighten the efficiency of the engineers themselves ; and he personally examined young aspirants until he became a marshal.

His untiring activity occupied itself with civil hardly less than with military reforms ; but to notice his various schemes for the benefit of his countrymen, or his services in connection with canals and harbours, would carry us too far. There are two of his efforts, however, which cannot be passed over : his protest against the Revocation

[6] Augoyat, i. 151.

of the Édict of Nantes, and his project of the "Dîme Royale."

His constant journeys throughout the length and breadth of France enabled him to judge better than almost any one else of the disastrous effects of the pressure put upon the Huguenots; and though a Roman Catholic himself, he could not refrain from making an appeal to the Government. Besides the loss of about 100,000 persons of all classes, who had carried into foreign countries the arts and manufactures which had hitherto drawn money to France; besides the transfer to the enemy's forces of about 20,000 excellent soldiers and sailors, and the heavy blow to French commerce; there remained behind a large body of disguised Huguenots, and of impoverished Catholics, "who say nothing, and who approve neither of forced conversions nor perhaps even of the present Government, which inflicts so much suffering on them," which would constitute a grave danger in case of invasion.

"It is no case for flattery," he says; "the interior of the kingdom is ruined, everything is suffering, and groaning; one has only to see and examine the heart of the provinces to find that it is even worse than I say. Instead of increasing the number of the faithful in the kingdom, compulsory conversions have produced only relapsed, impious, and sacrilegious persons, profaners of all we hold most sacred, and in fact a very poor edification to Catholics. Kings are, it is true, masters of the lives and property of their subjects, but never of their opinions, since the sentiments of the heart are beyond their power, and God alone can direct them as He pleases."

After pointing out the powerlessness of the country in

its divided state to carry on war successfully against the coalition that threatened it, he concluded :—

"On this account, looking to the importance of the matter, it appears that the King could not do better than put aside all other considerations as frivolous and unimportant by the side of this, and issue a declaration in whatever form may be best, in which His Majesty should state that having seen with sorrow the ill-success of the conversions, and the obstinacy with which most of the newly converted cling to the so-called reformed religion, notwithstanding their abjuration of it, and the hopes he had been led to entertain to the contrary, His Majesty, unwilling that any one should any longer be constrained in his religion, and desirous of providing, so far as rests with him, for the repose of his subjects, especially those of the so-called reformed religion, who for some time past have been obliged to profess themselves Catholics, after having committed the matter to God, to whom alone belongs the conversion of the heart, re-establishes the Edict of Nantes, purely and simply, on the same footing as it formerly was."

This memoir was written by Vauban in 1686, and he submitted it to Louvois and to Madame de Maintenon. But such advice was not likely to be well received at the Court of Louis XIV. Louvois returned the memoir to Vauban, recommending him to destroy it, and added :— "as I have never known you make such a blunder as you seem to have made in this memoir, I concluded that the air of Bazoches had clogged your wits, and that it would be a very good thing not to let you stay there much."

But his project of the Royal Tithe drew down on Vauban a heavier blow. It was in the two years of leisure that followed the peace of Ryswick (1697) that he brought this project into shape; but he had been gradually elaborating it for many years before. The extreme misery and destitution of the bulk of the population had pressed upon him in his constant journeys, as the extracts above given indicate; and at the same time he was struck by the comparatively scanty resources, both in men and money, which the State obtained at the price of all this suffering. The unequal incidence of the taxes, and the wasteful mode of collecting them, were the two main causes of this; the first he proposed to remedy by doing away with all class-exemptions, and the second, by substituting uniform taxes in produce or on income for arbitrarily assessed taxes on land. The Royal Tithe (not necessarily a tenth, but a proportion varying with the requirements of the State) was to be levied alike upon all the yield of land, upon rents, wages, pensions, or professional incomes, including the revenues of the clergy. Instead of a salt-tax varying in different provinces, and involving monstrous abuses, all the salt-mines were to be acquired by the Crown and the salt sold at an uniform rate. Customs duties on imports, and taxes on luxuries, together with the rents from the Crown lands, completed the scheme. Vauban did not content himself with throwing out crude suggestions; he laboriously gathered statistics, and worked out calculations to show the effect of the changes he proposed. The only objection, he concluded, to his system would be in "the self-interest, timidity, ignorance, and idleness, of those who might be set to examine it."

But the adverse influences which he thus anticipated, and was at no pains to conciliate, were too powerful for him.

"His book," says St. Simon, "was full of information and figures, all arranged with the utmost clearness, simplicity, and exactitude. But it had a grand fault. It described a course which, if followed, would have ruined an army of financiers, of clerks, of functionaries of all kinds; it would have forced them to live at their own expense, instead of at the expense of the people; and it would have sapped the foundations of those immense fortunes that are seen to grow up in such a short time. This was enough to cause its failure."

In 1699 Vauban sent a manuscript copy of his project to the King, and another to the minister Chamillard. How it was received is unknown, but at all events it did not stand in the way of his becoming a marshal three years afterwards. Probably it was ignored, for in 1704 he presented a second copy to the King, of which also no notice seems to have been taken. At length in 1706, Vauban, anxious to submit his ideas to a wider circle of readers, determined to print about three hundred copies for private circulation. The royal licence for printing which the law required, was in such a case certain to be refused. The copies were therefore printed secretly, and they were distributed by Vauban himself, who had just resigned the command of Dunkirk, and was living privately in Paris.

But in a few weeks the work was brought before the Privy Council and condemned. All copies of it were ordered to be seized and put in the pillory; and booksellers keeping or selling any were to be fined. The condemnation was secretly managed, so as to allow Vauban no opportunity of appealing to the King; and it took him altogether by surprise. His health was already much shaken, and this blow was too much for him. Profoundly dejected, he fell into a fever, and within a week he died, on the 30th of March, 1707. "I have lost a man very devoted to my person and to the State," was the comment of the *Grand Monarque* on hearing of his death: beyond this he showed no concern. The body was privately buried at Bazoches. The heart, a century afterwards, was brought to Paris by order of Napoleon, and deposited in the church of the Invalides.

CHAPTER III.

MONTALEMBERT.

THERE is one class of ancient monuments in behalf of which no Preservation Society ventures to interpose, though it is rich in historical associations. A girdle of ramparts is an irksome restraint to a growing town, and when it has lost its value in the present, it will not long be spared on account of its services in the past.

Within the last few years the fortifications of Antwerp have been pushed more than a mile outward, and its famous citadel, Alva's pride, which stood its last siege in 1832, has been swept away. Already new lines are to be seen springing up round Calais and St. Pierre, to replace the walls which detained Edward III. so long, and were strengthened to so little purpose by Henry VIII. The visitor to Namur will look in vain for the gate of St. Nicholas, or the demi-bastion of St. Roch, or the spot, so exactly described, where uncle Toby received his wound.

On all sides, in fact, bastions and curtains, hornworks and half-moons, are passing out of use and out of existence. A few green mounds on the hill-tops to right and left are all that meets the eye of a traveller as the railway carries him through the main line of defence of a great modern fortress. And upon a nearer view these forts will be

found to be very different in the details of their construction from the works which figured on the Shandy bowling-green.

The revolution which has taken place in artillery during the last five-and-twenty years has given the chief stimulus to these changes in fortification, but it did not give birth to them. More than fifty years ago the Prussian engineers at Coblentz and elsewhere set the example of discarding the bastioned lines which for three centuries had been the historical type of fortification, and resting their system of defence upon casemated forts. But the ideas which they put in practice had found expression some time before. The perfection to which Vauban had brought the art of attack, and the consequent abridgment of sieges, had led many men in the eighteenth century to cast about for some fresh means of restoring the balance between besieged and besieger. The Prussian engineers were thoroughly eclectic. They professed to draw freely from the whole literature of fortification, and were fond of tracing back their methods to Durer, and other German authors of old time. But it was unquestionably to a Frenchman, the Marquis de Montalembert, that they were most largely indebted. Like other prophets, he had for a long time little honour in his own country; but the war of 1870 taught the French many lessons, and broke down obstinate traditions. The defences both of the frontier and of the capital have had to be reconstituted on a gigantic scale, and at the end of a century, as a German writer says:—

"That great French engineer, to whom Germany raised the finest of monuments in the vast fortifications executed entirely in his spirit has now at length received from a later generation, in his own country, the recognition which he so well deserved."[1]

It is to be hoped that France will before long make amends in other ways to the indomitable man of whom she has hitherto been so strangely neglectful.

The literary monument of his own creation, the eleven quarto volumes of the *Fortification Perpendiculaire*, is deterrent by its mere bulk and costliness; and the work of condensing it has been left to a Prussian officer, General von Zastrow. No biography worthy of the name has been written of him. His nephew (the father of the celebrated Comte de Montalembert), who as a young man shared his labours, and wrote of him with warm attachment, and who afterwards, when serving in the English army, translated part of Carnot's treatise on the defence of fortresses, would have been his best biographer; but, owing, doubtless, to the circumstances of the last years of his life, no family memoirs of him have been published.

The facts of his history have to be gleaned chiefly from the incidental allusions in his own works, with slight help from the meagre notices of him written shortly after his death by his friend, the astronomer Lalande,[2] and by De Sales,[3] a literary man employed by the second Madame Montalembert.

We shall try to bring these facts together, and without entering into a detailed description of his systems, to give some account of the controversy to which his proposals gave rise.

---

[1] *Militär Wochenblatt.*
[2] "Magasin Encyclopédique," A.D. 1800.
[3] "Vie littéraire du General Montalembert." Republished in the sixth volume of De Sales's collected works.

Marc Réné, Marquis de Montalembert, was born at Angoulême on the 16th July, 1714. He came of an old Poitou family, a branch of which had settled in Angoumois. André de Montalembert, Comte d'Essé, who commanded the French troops sent to help the Scotch after the battle of Pinkie, was of the same stock. He fortified Leith after the then new Italian manner, with bastioned fronts. His gallant defence of Landrecies, and of Thérouanne, where he died upon the breach in 1553, were celebrated by Brantôme, and furnished his kinsman two centuries afterwards, with examples of the way in which places were defended in old time.

Montalembert himself entered the army in 1732. He was present at the brief siege of Kehl, in 1733, the first operation undertaken by the French in their war with the Emperor on behalf of Stanislas of Poland, the father-in-law of Louis XV. In February, 1734, he obtained a captain's commission in the cavalry regiment of Conti. He took part in the forcing of the lines of Ettlingen, and in the siege of Philipsburg. The latter gave him a good opportunity, before he was twenty years of age, of learning the art of sieges. The besieging army under Berwick was 60,000 strong, but it felt insecure, for Prince Eugene was at the head of the Imperial forces, and most complete lines of circumvallation were thrown up, lest he should interrupt the siege. The garrison was weak and passive, but the floods of the Rhine fought for it, and enabled the place to hold out for more than six weeks. Marshal Berwick was killed within the first fortnight by a shot from one of his own batteries, and D'Asfeld, who succeeded him, was a cautious commander, indisposed to risk assaults.

In 1741 came the war of the Austrian succession. Montalembert had not the good fortune to be with the French army which invaded Bohemia and escaladed Prague. His regiment formed part of the army of Marshal Maillebois, sent into Westphalia to watch the Dutch and Hanoverians. But in the summer of 1742 that army also was directed upon Bohemia, to relieve the troops under Marshals Broglie and Belleisle, who were by that time shut up and besieged in Prague. Maillebois did not consider himself strong enough to penetrate to Prague, but, after reaching the frontier of Bohemia, he turned southward to the Danube. He was soon afterwards superseded by Marshal Broglie, who had made his escape from Prague. Montalembert had now left his regiment to become the aide-de-camp of the Prince of Conti, and he remained attached to that Prince for the rest of the war. Conti had joined Marshal Maillebois without leave, after being refused permission to serve in Bohemia; but though placed in arrest by the king's order for a few days, he was allowed to continue with the army. He was at this time twenty-five years of age, and though he had made the campaign of 1733, he had as yet only earned himself a name in the scandalous chronicles of the time. He proved, however, a soldier not unworthy of the Condé blood, so far as his limited opportunities allowed.

Conti was in command, and Montalembert was present, at Deckendorf on the Danube, when it was attacked by the Austrians on the 27th May, 1743—the day of the Battle of Dettingen. A bridgehead had been made on the left bank of the Danube, and a line of communication formed from the bridgehead to the town, which had been prepared for defence and strengthened by four redoubts.

But these redoubts were abandoned by their garrisons after a few hours' cannonade, and Conti finding that his troops would be shut up in the town, withdrew them across the river. The loss of this important post, just above the junction of the Iser with the Danube, obliged Marshal Broglie to fall back, ultimately to the Rhine.

In 1744 the Prince of Conti received the command, in conjunction with the Infant, Don Philip, of the combined French and Spanish army which was to invade Italy. He raised a company of guards, of which he made Montalembert captain: all men of birth, magnificently dressed, according to Barbier. The Italian campaign opened brilliantly. Moving along the coast, the allied armies, about 45,000 strong, stormed the intrenchments on the heights east of Nice, and quickly reduced the fort of Montalban and the citadel of Villafranca. Then, recrossing the Var, they turned northward, and penetrated in several columns from Dauphiné into the valley of the Po. There was some severe fighting in the passes, especially above Castel Delfino, where the advanced guard of the column engaged was commanded by Chévert, the daring officer who had been the first man to enter Prague and the last man to leave it. Montalembert has referred to this engagement, of which he was an eye-witness, as an instance of the importance of good flank defence for the ditches of fieldworks. The French troops, after losing many men, got into the ditch of one of the redoubts, and finding themselves safer there than they would have been in retiring they stayed there until they had managed to pick away the rocky scarp with their bayonets, when, mounting on one another's shoulders, they carried the work.

Meanwhile the main column under Conti himself had

forced "the barricades" at the head of the valley of the Stura. The rock fortress of Démont, upon which the King of Sardinia had lavished vast sums of money, was taken in a week; and the army pushed onward to Coni. There were great rejoicings in France, and Louis XV., at a supper immediately before his illness at Metz, drank to the health of "my cousin, the great Conti." But the tide of success turned. The Spaniards insisted upon attacking Coni on its strongest side, and Montalembert, whose special function it was to follow the siege-works and report on their progress to the prince, early predicted that the attack must fail. The King of Sardinia, who advanced to raise the siege was repulsed; but the rainy season came on, and after thirty-six days of open trenches it was found necessary to abandon the undertaking. The army retired upon France, demolishing the castle of Démont on the road.

Montalembert was made a colonel in October, but he still continued on the staff of the Prince of Conti, who in the following year commanded the French army on the Rhine, and won some credit by his skilful retirement across that river at Nordheim.

In 1746, the electors of the Empire having declared themselves neutral, Conti was ordered with the greater part of his troops to the Low Countries, to reduce the fortresses of Hainault. Mons was taken in sixteen days, St. Ghislain in four days. Charleroi was next besieged, and this place, which had resisted Vauban for twenty-six days in 1693, now held out only five days. After its surrender Conti joined Marshal Saxe, but the two could not work together. Conti was eager to attack the enemy, and as a prince of the blood he claimed to be under no other officer's orders. Saxe insisted on his right to decide as generalissimo whether to give battle or not, and preferred to manœuvre. Conti, who was generally blamed, was recalled to Paris, and could never obtain another command. He lived for thirty years, a disappointed man, opposed to the foreign policy of the court, and supporting the parliament of Paris, and got the nick-name of "M. l'Avocat" from the king in place of "le grand Conti."

Montalembert probably remained with the army, for he is said to have been present at the siege and capture of Namur, by which Marshal Saxe closed the campaign. But he does not appear to have taken any part in the campaign of 1747; and though he refers to the siege of Maestricht, and to the great siege of Berg-op-Zoom, it is not as an eye-witness. He turned his attention to practical science, and read the first of what proved to be a long series of memoirs before the Academy of Sciences, which at this time (1747) admitted him as an associate. The subject which especially engaged him was the improvement of cast-iron ordnance, to supersede the costly bronze guns then generally used, both on board ships and on shore. He established ironworks at Ruelle in 1750, which proved most useful in supplying guns for the navy, though to the detriment of his own fortune. Buffon in his memoir on the improvement of naval ordnance, says of him :—

"I should be more flattered by his approval than by that of any one else; for I know no one who understands this subject better. If the wealth of ability which in this kingdom is flung aside as if with contempt were all gathered up, we should soon be the most flourishing of nations and the richest of people."

In 1752, Montalembert was made Lieutenant-General of

the provinces of Saintonge and Angoumois, and also obtained the much coveted honour of a commission in the company of light horse of the Guard. In 1757, the Seven Years' War having begun, a French army under Marshal D'Estrées marched upon Hanover, and defeated the Hanoverian troops at Hastenbeck. The Duc de Richelieu soon afterwards obtained the command, and on the 8th September concluded with the Duke of Cumberland the convention of Closter-Zeven. Montalembert, who was with this army, was sent by Richelieu at the end of October to the headquarters of the Swedes, who had invaded Prussian Pomerania. In the following spring he was made a brigadier, and was regularly appointed the French Commissioner with the Swedish army; and he remained with it throughout that year. His correspondence with the French ministers and ambassadors during this period, and for some years afterwards, was published in 1777,[4] and is of interest not only for his own history, but for the general history of the war. "A very intelligent, industrious, observant man," says Carlyle; "still amusing to read if one were idler."[5] Not much was to be made of the Swedes, however. The senate, many of whose members were in French pay, had joined in the league against Frederic, in spite of the strong opposition of their king, Adolphus; and senators and generals alike were chiefly eager to shift from their own shoulders the responsibility for any failures that might occur. Councils of war were called at every step. Montalembert writes:—

[4] "Correspondance de M. le Marquis de Montalembert, &c., &c. pendant les campagnes de 1757, 58, 59, 60 et 61." 3 tomes, Londres, 1777.
[5] "Frederic the Great," v. 275.

MONTALEMBERT. 107

"The general dare not decide himself; there must be unanimous assent. You will see what an awkward position I am in when I have to oppose openly the decision of a council of war, for I cannot show the soundness of my own opinion without proving that every one else is wrong. Still I cannot leave them to do what I think mischievous and I employ the utmost tact. As to the point itself I bate nothing, but I am full of consideration for all that touches that self-esteem which is carried to such a pitch in this people." [6]

In 1757 the Swedes, as one of their own senators said, "went into the enemy's country like foxes, and came out of it like hares." In 1758 they moved irresolutely, at one time eastward to join the Russians, at another time westward to co-operate with the French or the Austrians; effecting nothing except the feeding of their troops in the enemy's territory.

Montalembert tried in vain to induce the Swedish and Russian governments to undertake conjointly the siege of Stettin, which would furnish them with an admirable base for further operations. Failing in that, he endeavoured with no better success to persuade the Swedish general to make a dash at it with his own army only. He drew up a detailed project for escalading it with 16,000 men, proposing himself to act as the guide for the columns of the main attack. He afterwards appealed (in a footnote to the correspondence) to Loudon's successful assault of Schweidnitz in 1761, in justification of his scheme.

During the winter of 1757, when the Swedes had to retire before the Prussians into Swedish Pomerania, and were ultimately blockaded in Stralsund, he found some

[6] 18th Sept. '58.

work to do as an engineer. He put Anclam, on the Peene, into a state of defence; and at Stralsund he furnished a project for a retrenchment which he says the engineer of the place admitted to be better than his own, but which was not carried out. His predilection for work of this kind was already marked. He mentions that in 1757, judging from plans sent to him that the fortress of Louisburg in Canada was exposed to danger from some high ground near it, he sent a design for some advanced works to occupy it to the minister of marine. He was told in reply that Louisburg was safe enough, and nothing better could be wished than that the English should lay siege to it. They did so, and took it, the key of Canada, next year. At the end of 1758, after a visit to Stockholm, which as usual drew from him loud complaints of his sufferings at sea, Montalembert went to Paris, to report verbally to the ministers. It was decided that for the next campaign he should be attached to the Russian army. Leaving Paris on the 20th May, 1759, he travelled rapidly by Vienna to St. Petersburg, for, "I shall be in despair," he writes, "if a battle take place before I join the army." He reached St. Petersburg on the 4th July, but he had to wait there until the end of the month, in order to be presented to the empress. Consequently he did not arrive at the Russian headquarters till the 20th August, a week after the battle of Kunersdorf. He pressed the Russian general to improve his victory, but to no purpose. Soltikof roundly declared that his men had done their share of the fighting, and it was now the turn of the Austrians. When he had at length agreed to co-operate with them in Silesia, Marshal Daun found it necessary to march into Saxony, and Soltikof, angry and alarmed, was with difficulty restrained by Montalembert from retiring at once upon Poland. He was ultimately persuaded to remain in Silesia, inactive, but still a menace to the Prussians, until the middle of October, when he went into winter-quarters.

During the winter Montalembert continued indefatigably to advocate the siege of Stettin, and tried hard to remove the exaggerated impression of its strength which the Russians entertained. But he was once more disappointed. It was decided to open the campaign by a joint attack of the Russians and Austrians on Breslau. But the Russians were too late; Prince Henry of Prussia had relieved Breslau and forced the Austrians to retire. Soltikof, always suspicious of being made the catspaw of Austria, and afraid that he would have all the Prussian forces on his hands, held cautiously aloof for some time. Montalembert recommended a Russian advance on Berlin, to draw Frederic out of Silesia, and so divide the Prussian armies; and after an alternative scheme for the siege of Glogau had fallen through, his proposal was adopted. While the main Russian army halted upon the Oder, a corps under Czernichef pushed rapidly forward upon Berlin. Montalembert fortunately accompanied this corps, and when Czernichef, meeting with unexpected resistance, was on the point of retiring, he persuaded him to wait for the Austrian general, Lacy, who was on the march to join him. Lacy came up next day; the few Prussian troops in the capital fell back, and the raiders had just time to gather their rich plunder before the approach of the king obliged them to make off. Czernichef rejoined the main army by forced marches, and the Russians then retreated beyond the Vistula.

This successful stroke closed Montalembert's connection with them. He returned to France for the winter, and in February, 1761, he was made *maréchal de camp*. In April he was sent to Brittany to take command of a force intended for the relief of Belleisle, at that time besieged by the English. The relief proved to be impracticable, on account of the English fleet, and at the end of May he was appointed commandant of the isle of Oleron, opposite Rochefort, which it was thought would be next attacked.

This gave him a good opportunity of trying his hand as an engineer. He formed an extensive intrenched camp in advance of the citadel, with field-works of his own design. Its strength was not put to the test, but the skill and energy of the designer were much applauded. "All the world sings your praises," the minister of war wrote to him at the end of the year.

He was anxious to return to his old post with the Russian army for the next campaign; but the death of the Empress Elizabeth in January, 1762, led to the withdrawal of Russia from the coalition against Frederick. Before long France followed Russia's example, the Seven Years' War came to an end, and Montalembert saw no further service in the field. He was now forty-eight years old. He had taken part, as he reckons, in fifteen campaigns, and been present at nine sieges.

The reforms in fortification to which his life was henceforward to be devoted had already begun to take shape in his mind. In the beginning of 1761, immediately after his return to France, he had prepared a prospectus of a work on "perpendicular fortification," which he asked authority to publish. The Minister of War referred the prospectus to two officers of Engineers—General Filley and Colonel Fourcroy. The former, while expressing his agreement in principle with Montalembert, thought it undesirable that so useful a work should be made common property by publication. Fourcroy was of the opposite opinion.

"All sorts of people who have aspired to become the authors of new systems of fortification have hitherto failed," he said, "from not knowing wherein precisely the problem consists. Though it is of no use to enlighten people who have not made this study their profession, yet neither can there, in my opinion, be the least danger in letting the public know the combinations and assertions which may be the fruit of M. de Montalembert's labour. If the work is good, it is at best the fancy of a *savant* which cannot be turned to account either for or against the state."[7]

The Duc de Choiseul, however, adopted General Filley's view, and appealed to Montalembert's patriotism to reserve his ideas entirely for the king's service. Otherwise, he wrote, "if we profit by them, neighbouring powers will have the same advantage against us, and it will be just as it was with the invention of gunpowder."

Consequently it was not till 1776, when Louis XVI. had succeeded to the throne, and Maurepas, recalled to court after his long disgrace, had become chief minister, that the first volume of "La Fortification Perpendiculaire" was given to the world. A second followed in 1777, and a third and fourth in 1778. These four volumes, largely as they were afterwards supplemented, were regarded by Montalembert at the time as a complete work, fully

[7] Augoyat, "Aperçu sur les Ingénieurs," &c. ii. 541.

developing his principles alike for fortresses, coast-defences, and field intrenchments.

In the preface to the third volume he sums up their contents.

"All the parts of the art of defence have been treated of successively; and we have shown how the value of all of them may be increased. The inherent vices of the bastioned systems have been pointed out; we have proved by authentic facts that the flanks in this method of fortifying were of no effect; we have given means of making places already built according to this method infinitely stronger than they are, and our new angular towers have raised these means to the highest degree of strength. We have shown that towers of this kind admit of the most general use. . . . . . Forts of various sizes, of various shapes—square, round, and triangular, and of various degrees of strength, have been described in the utmost detail. One of these forts, which we have named Fort Royal, has been proved impregnable, by the systematic attack we have worked out upon it.

"Our chief attention in all our studies has been directed to the fire of the defence. It is necessary that this should be *protected* and *abundant;* these are the conditions to be fulfilled in the art of defence. As the ramparts now in use satisfy neither one nor the other, we have had to seek some other mode of meeting them. We have employed arches, we have given a new shape to the embrasures which considerably diminishes their size, and lastly we have completely closed their openings by solid shutters, which open and shut readily.

"After having thus provided for protection, we have largely increased the quantity of our artillery fire by devising new carriages, which need only three men, instead of eight, for the service of each piece, and allow the pieces to be at intervals of nine feet, instead of eighteen feet, as is usual with batteries on the ramparts. We have also made our casemates in three tiers, so as to have in a given space six pieces instead of one. Such advantages are all the more valuable because they are entirely reserved for the besieged. Constructions of this kind are the work of peace; it is not in a besieger's power to build similar batteries. Restricted to fascines held in their place merely by pickets, he can mount behind these fragile epaulments only a single piece in that space of 'eighteen feet wherein we make use of six pieces perfectly protected; and thus it is that we become necessarily the stronger, and that the few acquire a decisive superiority over the many.

"This is our fundamental means of defence. We have employed it wherever we have thought it necessary, whether for outer or inner works; for however great the superiority which we have given to the former, we have deemed it right to be equally careful of the latter. Revetment walls sustaining the ramparts presented the greatest obstacles to this; a dry ditch was required behind the walls of the enceinte in order that they might be defended on the inside; so we have detached and casemated them, and, built in this way, and strongly guarded as they are by our protected fire, their destruction has become impossible for the enemy.

"It is by following a road so different from that which is commonly taken that we have built up our new systems. As regards new fortresses it has set us free from those contorted outlines which are alike costly and unfavourable for defence. Simple angular enceintes are preferable.

Seen both inside and outside, they offer no part to the enemy where he can get cover; all is exposed to the fire of the besieged, which is everywhere very superior to that of the besieger, and the latter cannot advance a step without being hit from all sides."

He goes on to speak of his designs for coast batteries and field intrenchments which form the main subjects of his third and fourth volumes. In addition to these his work contains lengthy digressions, in which he reviews and criticizes in a lively and pungent style the wars of Louis XIV., and the early campaigns in which he had himself taken part. "His age, his rank, his experience, his ability, and his credit at court, allowed him," says Lalande, "to speak his mind at a time when no one else ventured to do so."

Without going deeply into details, we must explain a little more fully the nature and connection of his various proposals.

Gun-casemates are, as he himself indicates, the foundation of the whole structure. In themselves, of course, they were no novelty; they had been made use of to some extent from the very rise of modern fortification, and they had early assumed prominence in the hands of Albert Durer. But as the bastioned method of fortifying became more and more prevalent, and the mode of attacking fortresses of that type was in its turn systematized, gun-casemates went out of favour. Placed in the flanks of the bastions, and sheltered by orillons from the front, they could nevertheless be searched out, it was found, by shots made to glance off the curtain. In the early part of the seventeenth century De Ville wrote of them as obsolete:—

"Formerly men used to make vaults in the flanks, where the guns were placed under cover, and above them they made others for other guns: but this has gone out of fashion because of the great inconveniences which have been found to result in such fortresses; for after firing, the smoke so filled the vaults that it was impossible to stay in them, or to see one's way to load again, whatever air flues there might be, besides that the shock of the gun threw everything out; and the enemy firing into these low vaults, the splinters and fragments killed and wounded those within, and in a few rounds made a ruin of them; then those below being breached, those above gave way of themselves. On this account these vaults have been abandoned, and the low flanks are made open." [8]

Low flanks were altogether given up in the French school half a century afterwards in consequence of the introduction of the tenaille, in front of the curtain; and the defence of the ditch was derived wholly from the ramparts. By this time the bastions had grown in size and importance, and the curtains had declined until the latter had become little more than a means of keeping the flanks far enough apart to allow the shots from one or other of them to reach every part of the ditch. The whole *raison d'être* of the bastioned trace had come to lie in the exigencies of flank defence from high and open ramparts. But meanwhile this kind of defence had suffered greatly by the improvements in the vertical fire of shells and by the introduction of ricochet fire. Vauban himself in his later years had recurred to casemates for the flanking of his curtains, and gave it as one of the

[8] "Les Fortifications du Chevalier Antoine De Ville." Paris, 1628.

chief recommendations of his tower-bastions that "they have no occasion to fear ricochet or shells, which are the scourges of places now-a-days." While Vauban's successors, from some inexplicable prejudice, would not follow him at all in this direction, Montalembert went far beyond him, and laid it down broadly that "a flank protected by a bombproof roof is better than an open flank" (vi. 2). Casemates can be sunk to the level of the ditch, so that his reliance upon them set him free from the concern about "dead angles" which had made the bastioned trace a necessity; he was able to insist upon all its inconveniences, and to adopt other modes of trace instead of it. From the form of a bastion, shots aimed directly at one face or flank, enfilade or take in reverse the other face or flank; so that the bastions become a very focus of fire from all quarters. There is not room for good self-flanking retrenchments in the gorges of bastions, and it is impossible to flank them from the bastions next to them. Such are some of the defects which he points out. Again, owing to the interposition of a long curtain, which is itself useless for defence, the flanks forfeit the advantage of the full range of firearms. Their chief duty is to defend the faces of the bastions; how much better, then, to bring them up nearer to their work, and to place them back to back in the middle of the front! They would be more sheltered there, and the fronts could be made longer and fewer, which would save both men and money. In this way he was led to his polygonal trace. But though it is upon this line of thought that his ideas have proved most fruitful, he himself rather leaned to another solution of the problem. Engineers have tried to reinforce bastioned fronts by the addition of ravelins, which in fact convert them to an angular or tenaille trace, the bastion-faces and ravelins defending one another, and their salients becoming necessarily the points of attack. This arrangement is quite to Montalembert's mind, for it is one of his axioms that the longer a flank is, the better it is (vi. 2). But the ravelins, lying beyond the main ditch, are too far from the body of the place, and their communication with it is too insecure, for a stout defence. It would be much better, instead of trusting to outworks, to have an angular enceinte, its faces defending one another and strictly perpendicular to one another (whence the title of his work, "Perpendicular Fortification"). The salients, or star-points of such an enceinte, can be readily provided with successive retrenchments.

"Increase and secure the fire of the flanks; oppose to the enemy powerful obstacles inside the main ditch; and if this fire and these obstacles are what they ought to be, you may be easy about the fate of places." This, he says, is the base of his theory, the most salutary principle of fortification (i. 131).

The construction of the casemates flanking the ditches, which would be much the same whether they were placed in the re-entering angles, as in the angular trace, or in the middle of a front, as in the polygonal trace, was a question of prime importance, which he had carefully studied, and minutely described.

There were two things to be guarded against: the smoke from their own guns, and the shots from the enemy's breaching batteries. As regards the latter, it was Montalembert's principle that the best means of preserving the casemates was to enable them to overpower the batteries: "there are no walls stronger than those against which one cannot fire" (i. 135). As regards the smoke, it was customary to point to some trials of Vauban's tower-bastions (which seem by-the-bye, to have been contradicted by later experiments at Neuf-Brisach in 1799), as a proof that casemates soon become unendurable. To this Montalembert opposed the experience with some new casemates at Olmutz, which he had himself visited, and he pointed out the faults in the construction of the Neuf-Brisach casemates. He took care to provide ample means of ventilation in his own designs.

"I flatter myself," he says, "that I have succeeded in massing in a small space the most powerful fire conceivable, of artillery and musketry, with such numerous openings that one will breathe as well in them as in the open air; the broadside of a three-decker would not deliver such a fire as my casemated works."

He had taken many hints from naval architecture, and the block of casemates which he gives as an illustration may in fact be described as a stone man-of-war, moored in the middle of the main ditch. Each flank or broadside has two tiers of gun-rooms, and three tiers for musketry, besides the open batteries on the top. The guns in the gun-rooms are placed only eight feet apart, which is more, he says, than is found necessary on board ships of war; their front walls are very thick, so that the gunports are most inconveniently long and narrow. He calls this structure a "casemated caponnière," borrowing a term which at that time especially applied in France to the covered roadway across the main ditch in the middle of a front.[7] This caponnière is intended, of

---

[7] This roadway was commonly open overhead, and covered only

course, for the defence of one of the straight fronts of his polygonal trace, but its flanks correspond generally to the flanks in the angular trace.

But it was by no means only for the flanking of ditches that Montalembert looked to casemates. He employed them to supplement the frontal fire from the ramparts, by detaching and arcading the escarp walls. This gave him one, or sometimes two, tiers of gun-rooms, the guns of which, it is true, could only fire indirectly upon the country, but were much better sheltered, and at much closer intervals, than those upon the rampart behind them. He employed them also for retrenchments and keeps, especially in the form of the angular towers which were his peculiar pride. These towers had a star-shaped base, upon the principles of his angular trace, and loop-holed for musketry, so as to be self-flanking; but above this they were circular, their outer wall being carried on arches across the re-entering angles of the base, and pierced with gun-ports. They varied widely in size, some being only twelve yards in diameter, with two or three gun-floors for twelve guns each, while others were very much larger at the base, and were carried up, as it were, in terraces. The towers and the casemated escarp walls could be applied without complete reconstruction to separate

by earthen glacis at the sides; but the name is said to be derived from *caponera* (Spanish), a coop, and in Fäsch's Kriegs-Lexikon (Dresden, 1735), it is described as a covered way, arched over with stone, or with a roof of timber with earth thrown on to it, and provided with a palisaded parapet on both sides. Fäsch gives a sketch of a half-caponnière made by the Swedes at Stralsund during the siege of 1715, as a keep to a place of arms; and it was perhaps his service with the Swedish army which led Montalembert to adopt the term.

existing fortresses of bastioned trace; and Montalembert, eager to see his ideas put in practice even partially, made much of the strength which would be attainable in this way.

The angular trace, though applicable on quite a small scale to towers, was not well suited to earthen forts. It gave a star of at least twelve points, as the salients could not be made less than 60°, and each of the faces ought to be at least 100 yards long. It is true, as he points out, that such a work would not occupy more space than a square bastioned fort, which requires an area of nearly 600 yards square, and a garrison of 2400 infantry, and costs two millions of livres. "But how many eminences that it is advantageous to occupy have not that extent," or if they have, cannot be occupied at so large an expenditure of men and money? Montalembert especially insists (and it is one of his chief titles to distinction as a pioneer in military science) upon the importance of making large use of detached works, especially to secure naval arsenals from bombardment.

"A regular siege," he says, "is not to be feared in the case of the ports of the kingdom. The large forces that one would be able to bring together while it was going on would not allow an enemy to attempt such an undertaking. The chief thing to be considered, therefore, is how to keep him for a certain time beyond bombarding range." (iii. 38.)

To obtain works suitable for such use, and capable of ready adaptation to every variety of site, he turned to his polygonal or caponier trace, though he adopted that also sometimes for extensive enceintes on account of its simplicity and cheapness. Marshal Saxe—"one of those men of whom nature is too sparing, those men of genius who are creators in every subject they take in hand"—had fortified his camp before Maestricht in 1747, with redoubts about 35 yards square. Taking these as his starting-point, Montalembert, who at Castel Delfino and elsewhere had noted the weakness of works with unflanked ditches, shows how they might be strengthened by a palisade in the ditch, flanked by a small stockade caponier in the middle of each face.

He follows up this by design after design for square forts of increasing size and complexity, until he comes at last to the largest and most elaborated of them, a work nearly 400 yards square, which he names "Fort Royal."

In these designs he gives full play to his exuberant fertility of invention. "We have made variations on purpose," he says, "one fresh idea gives birth to another, and instruction cannot fail to gain thereby." Elsewhere he quotes approvingly General Lloyd's remark, that the "uniformity which exists in the works of engineers proves that they have not a spark of that genius which varies infinitely, and which forms new combinations to meet the new circumstances which must and do occur (x. 7)." But he himself seems hardly sufficiently alive to the distinction between such adaptation to special conditions and the unsorted profusion of ideas poured forth by men of more imagination than judgment.

In Fort Royal, which one may fairly call the quintessence of his projects, we find the casemated caponiers, the powerful flanking batteries, the detached walls, and the angular towers all brought together. Each of the four corners presented to an enemy, first, a detached wall with two tiers of gun-casemates, secondly, an earthen rampart, and lastly some casemated barracks forming an incomplete square, in the middle of which was an angular tower of four stories.

The central portion of each front was retired and formed a casemated cavalier, cut off by a small ditch from the rest of the work. The main ditch was flanked by caponiers, one in the middle of each front, having three tiers of gun casemates, with nine guns in each tier on either flank. They were detached from the body of the work, and were themselves flanked by similar batteries at the extremities of the casemated cavaliers.

Montalembert had been careful at first to hide his casemates from the enemy's view as long as possible, (excepting the upper floors of the towers); and on this account, though he held outworks cheap, he had a high rampart forming a continuous envelope or couvreface, outside the main ditch. Ricochet fire, he maintained, would not make breaches, and for direct fire it would be necessary for the enemy to make batteries on the narrow and exposed surface of the couvreface. But he soon grew bolder. He convinced himself that if these batteries ould be overpowered by his casemates, so also might others at a greater distance; that his maxim " one cannot knock down a wall at which one is unable to fire " admitted of wider application; and that a besieger should not be allowed ever to reach the glacis. With this view therefore, in his second volume (to which Fort Royal elongs) while still retaining the couvreface, he added a third tier to his principal casemated batteries, which looked over it upon the country. The direct fire from this highest tier would be supplemented by indirect fire from the tiers below it, and from the upper tier of the detached wall.

"There is not a single point," he says, "in the whole parallel where the enemy can establish his batteries either to right or left of the capital of the work attacked, which will not be under fire from more than a hundred guns" (ii. 238), apart from those that may be mounted on the ramparts. Hence he concludes that the making of batteries and the advance of saps will be alike impossible, and a fort of this kind, with a garrison of 1200 or 1500 men, is *impregnable*, while its cost (as he estimates) is less than that of a square bastioned fort, reckoned to hold out for a fortnight or at most three weeks. "Since it is impossible here to pretend that the besieger's artillery can be brought into action, it is absurd to dwell upon the destruction of embrasures and the splinters of stone." (iii. 170).

Objection might be made to the enormous armament required for the numerous gun-casemates, for in fact there are more than 2200 gun-ports altogether in Fort Royal. But as all four sides cannot be attacked at once, nor all the batteries come into play together, "300 to 400 iron guns will be more than is needed for the besieged to frustrate all the efforts of the besiegers, and that number might be reduced by one-third without endangering the place at all." (iii. 167). These iron guns, as he had himself proved at his own iron-works, could be made at less than one-tenth of the cost of the brass guns commonly used.

In addition to his square forts, he gave a variety of designs for triangular and circular forts; the former as cheaper and needing fewer men, the latter as best suited to certain sites. The circular forts had unflanked ditches, but they had casemated towers as keeps; and he had come by this time to expose his casemates so recklessly, and to trust to them so fully, that in some cases the earthen rampart merely screened the base of the tower, which rose nearly fifty feet above it; and with the help of a very gentle exterior slope to the rampart the ditch could be directly defended from the highest floor of the tower.

He illustrated the application of his different types of forts by a project of remarkable boldness and originality for the fortification of Cherbourg. Upon the heights to the south he provided two chains of works, the outer consisting of forts—six square, one triangular, and one circular—at a distance of about two miles from the place, and rather more than one mile apart; the inner consisting of small redoubts, as links between the outer line and the place. As a long siege was unlikely, he proposed a simple polygonal enceinte, flanked by caponiers which, instead of being detached, formed part of the general casemated escarp-wall.

The means to which Montalembert looked so confidently for holding in check land atttacks had obviously more to recommend them against atacks from seaward. The common impression that ships are more than a match for land batteries was due, he declared, to the faulty construction of the latter, which had commonly too few pieces and were seen into from the ships' tops.

"In general, barbette batteries are not suitable on the coast; one should only use them on flat shores, where ships cannot approach within musket-range, and in cases where the whole armament consists of a few heavy pieces which are required to fire in all directions; but this kind of battery is not capable of defence when the depth of water allows large vessels to come close. . . . . Every battery that can be approached within 300 yards should be built entirely of masonry and casemated, to oppose more resistance to the artillery brought against it, to prevent its fire being silenced by the musketry of the ships, and to allow the pieces to be nearer together. Such batteries should be in several tiers, like the vessels with which they have to deal, the pieces should be at similar intervals, there should be well protected musketry-fire if they can be approached within musketry-range, and then it will be the vessels and not the batteries, that will have anything to fear." (iii. 76, &c.)

He gave some designs for casemated coast batteries in his work, but he soon had an opportunity of giving a practical, and therefore more impressive illustration of his principles, and of bringing the smoke prejudice at all events to a test. The fort on the Isle of Aix, which covers Rochefort roads, had been taken and destroyed by the English in 1757. A project for a new fort had been prepared by General Filley, but the estimated cost was twenty millions of livres, and it was not carried out. In the beginning of 1778, by Maurepas' favour, Montalembert was invited to make a fresh project. He designed a masonry fort accordingly, but before the end of the year, war had broke out with England, and something was demanded that could be built more quickly. Resolved not to miss his opportunity, he designed a wooden fort as a temporary makeshift. The timber was put together under his direction at Rochefort, and was then taken over to the island, and erected there. The fort mounted fifty-six heavy guns in two tiers of casemates—sixteen in the lower, forty in the upper tier—and eighteen light guns on the platform above; and there was a low earthen battery of sixty-two pieces attached to it.

The Corps of Engineers did not hide their disgust at this intrusion of an amateur into their province. The general commanding in that part of France, the Marquis de Voyer-d'Argenson, wrote to the Minister of War, urging, on the strength of his engineers' representations, that Montalembert should not be allowed to build such a work. "It is the project of M. de Maurepas, and consequently of the King," replied the Minister of War; "so I do not advise you, my dear general, to oppose it." Montalembert tried to draw some expression of approval from one of the Engineer officers who were employed under him during the construction, and asked him if he did not think one novel feature an improvement; but he could only get the answer that "he did not think at all." Any such approval would have been dangerous, for the man who practically ruled the Corps of Engineers was Fourcroy, one of the two officers to whom Montalembert's original prospectus was referred, but who had by this time passed from an attitude of contemptuous indifference to one of vehement antagonism. He rashly predicted that the fort of Aix would prove a complete failure; not only would the "pestilential smoke" make the guns unworkable, but the concussion would bring the whole structure down.

To put these questions out of doubt, it was decided that the work should be tested by rapid and continuous firing from all the guns, as though it were actually engaged with an enemy. The trial took place on the 7th October, 1781. The guns were fired first successively, then independently, and lastly by salvoes of tiers, and a general salvo of the whole. The number of shots amounted to 523, fired in two hours.

"It was unanimously agreed," wrote Voyer d'Argenson to the Minister of War, "that it would have been useless to prolong it . . . the smoke of the guns in the first and second (covered) battery is, according to the acknowledgment of the naval officers themselves, much less in quantity, and less annoying than it is in the batteries between decks on board ships. The guns are mounted on very ingenious carriages. . . . It must be allowed that this new theory accumulates a greater mass of fire in a smaller space than the system of fortification in vogue, and that it seems to require fewer troops for its defence. The work is substantial and good to judge from this trial." (v. 41).

It is no doubt true, as his opponents at that time maintained, that scattered earthen batteries, when they can be suitably placed at a moderate height above the water, are better than casemated batteries; will do more damage, and will suffer less. But suitable sites are rarely to be met with where they are most needed, and Montalembert's expectation has been fully borne out, that the fort of Aix "will always be reckoned the father of the numerous posterity which the future has in store for it." (vi. 29).

The construction of the earthen battery which Montalembert found it desirable to attach to his wooden fort, led him to give more consideration than he had hitherto done to the mounting of guns on open ramparts.

Vallière and Gribeauval had lately introduced higher carriages for fortress artillery, in order to avoid weakening the parapets by deep embrasures, and to make it easier to shift the guns quickly from one point to another. But Montalembert, seeing that ricochet fire was more to be feared than frontal, thought it better to keep the guns low. He placed his pieces by pairs between traverses and covered them from the front by shields of masonry with wooden port-frames for the ports; the general arrangement corresponding, except as regards the material of the shield, to one of our latest types of coast batteries. He contrived a strong and simple traversing platform with a front pivot, which, even with rapid firing, could be worked by three men.

To describe his gun-carriages and show their suitability for ordnance of all classes, he published a fifth volume in 1784, which he supplemented by a sixth in 1786.

"I have given my methods for the fortification of places," he says in his preface, "but I have not given those which relate to artillery. This part is no less susceptible of improvement. On the construction of embrasures and carriages depends the effectiveness of this, the most potent, means of defence." (v. 1).

But he had also another object in again taking up his pen, and the substance of these volumes, as of those that succeeded them, is controversial. The smouldering hostility of the Corps of Engineers had begun to show itself openly. In 1780, Major Grenier, a retired officer of that corps, had addressed a memoir to the Minister of War, entitled "Observations on the principal systems of

M. le Marquis de Montalembert," in which he attempted to estimate the value of these systems by working out the journal of an imaginary attack upon them, according to the rules laid down by Cormontaingne. Montalembert's sixth volume mainly consisted of this memoir, and of a reply to it, paragraph by paragraph, in the tedious style of old-fashioned polemics. But before fighting his own battle, he skilfully prepared his readers' minds to sympathize with him against a narrow *esprit de corps*, by recounting Cormontaingne's treatment of Bélidor. No more signal instance could well be found, of rare talent hampered by unworthy professional jealousy, than is afforded by Bélidor's whole career. He began life as a soldier, and served in Villars' last campaign in the Low Countries, but he soon devoted himself to science, and was made professor of mathematics at the artillery school at La Fére. After twenty years of invaluable work there, which bore fruit in all branches of military science, but especially in mine warfare, he was driven away by the outcry of the leading artillery officers, because he proved that they put too much powder into their guns. But he had a steady friend in Marshal Belleisle, and in 1740, when Belleisle assumed command at Metz, where large works of fortification were in progress, he sent for Bélidor, in order that a new system of his contrivance, which had been much fancied at Court, might be thoroughly discussed with the engineers there. Of these, Cormontaingne, though not the chief, was the leading spirit. He had prepared the projects which were in course of execution, and he was now called upon to write a memoir upon the new system. An able man, but bitter and opinionated, he has characterized as ignorant, vicious, and unpatriotic, those of his superiors who in any way overruled or slighted him. He was not likely to show much respect to a volunteer who came to teach him his business. He drew up an adverse memoir, making short work of the system, but he begged that Belleisle would keep it to himself.

"It was right that he should be fully informed upon the subject, charged as he was to report on it to the Court. As to the Sieur Bélidor, from whom the memoir was to be withheld, Cormontaingne considered that it was useless for him to see it on every account, as his duties did not call him in any way to have dealings of this kind with the Engineers, and to show it to him would only be to set him upon idle modifications of his new system, which would enable him to give yet more trouble to those who might have to reply to him." Belleisle found fault with his tone: "it was not in that style that matters affecting the King's service should be treated." But Cormontaingne justified himself on the ground that Bélidor's "conduct and proceedings were extremely culpable ; for in the previous winter he had presented himself at the houses of most of the lords of the Court, with this precious specimen of fortification in his hand, in order to gain patrons, and had found means of getting access even to the King himself, from whom he had obtained a pension of 1200 livres a year, without having received the support of a single engineer upon a subject which strictly concerns them, and much more, without the concurrence of their minister. He had the audacity to offer an application of this absurd system, as an amendment on the project drawn up at Metz for the heights of Montigny, a place of which he had only seen a painting." (v. 188).

Montalembert had served with Bélidor in Italy, and was his colleague in the Academy of Sciences. Besides a fellow-feeling with him as regards this professional arrogance, there were some points of likeness between their systems, such as the use of loop-holed detached walls, which Cormontaingne altogether condemned. The exaltation of Cormontaingne's own system—for although the Corps of Engineers piques itself on its exclusive adherence to the principles of Vauban, " each individual has none the less his own particular system—his cherished child, the most perfect thing that has appeared " (v. 159)—was an additional provocation to Montalembert.

But he had a further quarrel with the gauge of merit devised and employed by Cormontaingne—the method of imaginary attacks.

" It is not based on data that can be accepted ; it is in no respect conclusive, but altogether arbitrary. In Cormontaingne's attack upon Bélidor's system the garrison is perfectly passive, their artillery is without powder or ball, and while everything is asleep in the place, he advances with a boldness which would cost him dear if he had to overcome all the obstacles the besieged could oppose to him. Under such conditions as these Gibraltar would have been taken in five or six nights." (v. 201).

Instead of indulging in "arbitrary estimates of time," and calculating the advance night by night, comparisons of different systems should be based upon the obstacles to be overcome, and the fire to be subdued in each case.

Cormontaingne and his method soon found a champion. In March, 1785, Fourcroy presented to the Academy of Sciences a memoir on " Marshal Vauban's method of discussing questions of fortification." While anxious to get the benefit of Vauban's high authority on behalf of imaginary attacks, it was for Cormontaingne, " our modern Vauban," that he claimed the glory of being "the first to give fortification a solid theory which was wanting to it, and which placed it in the rank of exact sciences." "He was to our art,"he says, "what Newton was to physics."

Fourcroy had inherited the mantle of Cormontaingne, and enjoyed a double portion of his spirit.

"Bon ingénieur, l'esprit délié et net, s'énoncant et se présentant bien, le coup d'œil bon ; actif, et d'un grand détail : souple, haut, doux, ferme quand il le juge à propos ; un peu trop de politique et d'ambition."

Such was his portrait as sketched by General Filley, his immediate superior, in the early part of the Seven Years' War.[8] As he rose, his harshness to those below him and his insinuating address with men in power became more marked. In 1776, as the director of fortifications personally attached to the minister of war, he became the real head of the Corps of Engineers, and governed it rigidly for the next twelve years. As Carnot complained to the National Assembly in 1789 :—

"In France the Engineer corps is in strange bondage to the will of a single chief who is at the elbow of the Minister of War. The individual officers of this corps are so dependent on that chief, and his influence is so great on the fate of individuals, that it would be a most dangerous thing for them to have an opinion differing from his. Those officers are forbidden to print anything bearing on their profession, unless it has been submitted to the minister, or in other words to the chief he has given them."

[8] Augoyat, "Spectateur Militaire," 1836.

He added that Fourcroy had earned among them the nickname of "the extinguisher," and having declared Cormontaingne's system to be the best possible, did his utmost to proscribe any would-be rival. The whole of the professional manuscripts left by Cormontaingne had been entrusted to him, and from them he had compiled a series of guides to right thinking on questions of military engineering, which were privately circulated throughout the corps for many years before they were published. He retouched and developed Cormontaingne's work in several places, but nowhere so much as on this particular point of imaginary attacks.

In preparing either for the defence or for the attack of a fortress some sort of estimate must be made of the probable duration of the siege. Vauban's very large experience, and the general similarity of the places he had to deal with, enabled him to judge with unusual accuracy how long a given place might be expected to hold out; and as a practical engineer, when fortifying he took account of the cost. "The utility of these works being so unquestionable," he says in one of his memoirs, "we must see whether the expense may not be such as to cause us to reject them."

Upon this foundation, so far unimpeachable, Cormontaingne built his method of criticism, resting upon the cost of construction and the strength attained, as measured by the time required for regular siege operations prudently pushed forward. Fourcroy crowned the edifice by combining the two elements of cheapness and strength, assuming them to be of uniform and of equal value.

"Since fortification is better in proportion as it produces greater effect at less expense, its merit, or actual value in war may be expressed by the sum of its effects divided by the cost of its construction, and hence all works, or fronts of fortification, compared with one another, will have values proportionate to the quotients obtained from these divisions, that is to say (to use a term in mechanics) to the moments of those works or fronts.

"It is the investigation of these moments of fortification that we term its analysis. The scale obtained from these moments forms the veritable touchstone of the art, the delicate balance which can never deceive us as to the merit of productions of fortification. It is the use of this touchstone in our hands which places our art in the rank of the positive sciences, which distinguishes and separates what is soldier-like, useful, and well established from the arbitrary ideas, chimerical closet speculations, conjectural advantages and idle promises of all those well meaning authors who treat of this subject without understanding it."

As to the data of these calculations, drawn from the experience of past sieges, he admits that the duration of those sieges has been largely influenced by many circumstances more or less foreign to the character of the fortification, and springing from the character of the besieged or besiegers.

"It would seem at first sight very difficult, therefore, to reduce defences so disproportionate in duration to a fair average. But the Engineer Officer, who is versed in observing and calculating, very easily distinguishes and determines each of the effects which have been produced by each different cause. He very easily discovers and demonstrates how much the mines prolonged the defence of Berg-op-Zoom beyond the real merit of the works of Coehorn. He knows how to assign what was due merely to the valour of the besieged or the unskilfulness of the besiegers during the fine defences of Grave, Landau, Lille, Aire, Béthune, Douai, Prague, &c."

These accidents, then, can be put on one side, and we can work on the hypothesis that "the besieged making all possible use of his artillery and musketry, both by day and night, in all the works which can bear upon the attack, holds his ground firmly in all the works attacked, up to the moment when he would run the risk of being stormed if he remained an hour longer. We do not admit in journals of this kind any of those resources of industry or resolution of which we have spoken as entering into all good defences, for such circumstances would make the duration of the resistance on any front uncertain. Consequently we make up this journal of attack simply by calculating the time proved to be necessary for the assailant to execute completely, with all the speed that sound rules and prudence allow, the approaches, parallels, lodgments, batteries, breaches, descents, bridges or parapets across ditches, &c., in spite of the continual fire which he will experience from the besieged."

If it were a question, he adds, not of examining the fortification of a place, but of estimating the stores and ammunition for its defence, all the resources of a brave and intelligent garrison would be taken into account, and the calculated duration of the defence would perhaps be doubled. It is a mistake, therefore, to say (as is sometimes done), that this analysis pretended to fix the term of actual resistance; though Napoleon's allusion to "that craze of Engineer officers, that a place can only hold out for so many days," no doubt conveys the general impression of it, and its broad effect on men's minds. Its real fault lay rather in its assumption that the active and the passive elements—the garrison and their works—could be reckoned with separately, and that it was unnecessary to consider how far the energy of the former would find help or hindrance from the character of the latter. Also, as Choumara has pointed out, it omitted to take into account the expenditure of the besieger and the drain upon his resources. The air of precision given to it by its mathematical form disguised the doubtfulness of the facts upon which it was founded; which became mere guesses in the case of novel and untried systems, and made it, as Montalembert said, "a saddle to fit any horse."

The memoir presented to the Academy of Sciences was published by Fourcroy in 1786, together with a sequel applying the method to Montalembert's fronts, in a volume entitled "Mémoires sur la Fortification Perpendiculaire, par plusieurs officiers du corps royal du génie," which contained also a revised memoir by Major Grenier. Fourcroy who was assisted in his portion of the work by Captain de Frescheville, explains that Montalembert's success in getting ministers of state and soldiers of authority to adopt some of his ideas has induced the writers to depart in his case from the silence which they have habitually and deliberately observed regarding such productions. For just as the Academy of Sciences has decided not to examine any more of the schemes submitted to it for squaring the circle or for perpetual motion, so it is with fortification.

"An experience of more than seventy years has shown the officers of the Royal Corps of Engineers that none of those men who, since the time of M. de Vauban have published new systems of fortification, that is to say, who have treated that art as if it could be the fruit of closet speculations, have been acquainted either with its principles or practice. . . . It has in fact convinced them that a new system of fortification is nowadays one of the distinctive marks of ignorance of the art."

Montalembert's strongest compositions were disposed of by imaginary attacks in about three weeks. The possibility of breaching the exposed upper tiers of casemates at long ranges was rightly insisted upon; and Grenier—whose part of the work is the more temperate and more thorough—suggests that mere loopholed walls, though hidden, might be breached by indirect fire, a suggestion which the Woolwich experiments in 1824 showed to be at any rate not impracticable. But as Montalembert justly complained, no account was taken of his powerful artillery armament in pushing forward the approaches to the glacis, and the breaching batteries were assumed to be made without hindrance from the flanking casemates, which they then promptly silenced.

"If the fire of the place causes the besieger to lose workmen and officers during these operations (saps, batteries, and lodgments), since it is an accident always foreseen, and substitutes are always in readiness, the works will none the less make the progress calculated, decided on, and enjoined the day before upon the officers in charge."

To meet such a mode of argument, Montalembert had an apt instance ready to hand in the late siege of Gibraltar, which after lasting more than three years had been raised in 1783, when the approaches of the besiegers were still half a mile from the works.

"The rampart of the front attacked in this case was of

no account—a long curtain, flanked by two small bastions, but having on its left a mole running far out into the sea, and on its right a mountain, both of them capable of receiving a numerous armament. This front in itself one of the worst that could be made, could not be so much as tested; and why? because the very superior fire of the besieged did not allow any batteries to be established within breaching distance." (vi. 204).

As regards the general question of gun-casemates, Fourcroy, ignoring the Aix experiment, falls back as usual on the authority of De Ville, and dwells at length upon the behaviour of powder-smoke and the "mephitic gas" which chemistry proves it to contain, and which all Montalembert's ventilating flues will not carry off. He speaks of the disturbing concussion of the guns, and the demoralization of the men in casemates:—"The feeling of terror which would take possession of the bravest soldiers when the smallest stone, falling under these resounding arches, would present to them the idea of a total collapse and of the difficulty of escape."

He asserts that Vauban's immediate pupils have assured him (what nevertheless has found little credence) that before his death, in 1707, "that great man regretted the expense he had caused the king for those fine but useless gun-casemates for Landau, Belfort, and Neuf-Brisach; and that on any future occasions he would not have built tower bastions."

But if gun-casemates are to be condemned, what becomes, he asks, of Montalembert's forts and traces, and of the whole of his four volumes, "a sumptuous edifice which is entirely based upon the excellence of that protected fire which we believe impossible?"

Finding it at length necessary, in commenting on Montalembert's fifth volume, to notice the experiment in the Isle of Aix, he declares that this was altogether illusory.

"That wooden fort corresponded rather to the lower decks of a man-of-war than to a casemated work; but the trial did not turn on what takes place as regards smoke in the lower decks of ships, but on what would probably take place in casemates, which have no resemblance to these lower decks.

"Besides the batteries opened into an inner court, while our doubts about the smoke have reference only to casemates and casemated revetments closed in rear, or backed by the earth of the ramparts."

His predictions, which led to that experiment, are ignored altogether, and his lame reply provoked the rejoinder made shortly afterwards by Carnot,—"Well, gentlemen, make casemates which like the fort of the Isle of Aix shall be ship's lower decks; the name is no matter, as long as the right thing is done."

As was to be expected, Fourcroy's general conclusion regarding the first five volumes of the "Fortification Perpendiculaire" was "that among all the ideas, all the traces, all the practical precepts on our art which are given in this work we find nothing new which would not, if carried out, be injurious to the King's service."

Montalembert's reply was held back for two years by the veto of the Minister of War. It was at length published in August, 1788, and forms his seventh volume. It confines itself to Fourcroy's portion of the Mémoires, which it combats at all points, with needless diffuseness and repetition and much quotation from the earlier volumes.

Among his various Engineer opponents he reckons Carnot, at that time a captain in the corps, on the strength of some expressions used by him in a panegyric on Vauban. This drew from Carnot a very cordial letter disavowing any allusion to him in those expressions, and assuring him of his own warm admiration and of the friendly feeling of the corps in general. Men were now becoming more free to speak, but nevertheless no other Engineer officer openly took part with Montalembert.

Outside the corps of Engineers, he was beginning by this time to meet with a good deal of recognition. Frederick the Great is said to have invited him to Berlin, and Gustavus III. to have offered to place him at the head of the Swedish Engineers. Spanish, Dutch, and Danish officers declared themselves in his favour. In Prussia gun-casemates had always held their ground to some extent; and Major Lindenau, in asking permission to publish a German translation of the "Fortification Perpendiculaire," in 1788, told Montalembert that the Prussian Engineers were by no means of the same mind as their French brethren. "Our corps is ruled by the orders of our kings, not the authority of project-makers; and these have put no restriction on the invention of new systems."

In Mirabeau's work on the Prussian monarchy, Montalembert is spoken of as a man, "who, whatever professional charlatanism may have insinuated, has contributed very ingenious and very novel ideas to the art of fortification."

Indeed this had long been perceived by French ministers themselves. He had been several times consulted about works in contemplation, and one minister had written to him with cynical frankness :—

"I see plainly all the advantages of this project upon your method; but I will not hide from you my unwillingness to have it carried out; for if I let things go on as they went on before my time I am responsible for nothing; I am not bound to do anything better; whereas if I introduced novelties, however good they might be, there would be so many people interested in abusing them, that there would be a public outcry raised against me, which it is always dangerous to provoke in posts such as I hold." (viii. 19.)

But the course of political events now began to promise well for revolutionists in fortification, as in other matters. Established authorities, and vested interests were giving way one after another. On the 11th August, 1789,—just a week after the memorable sitting in which so many feudal privileges and immunities were abolished by the Constituent Assembly,—Montalembert received a letter from Mirabeau:—

"The National Assembly is about to form a Military Committee: you must not only be a member of it, but you must further be appointed by it Inspector-General of the fortifications of the kingdom, and reporter on the changes to be carried out, the means of putting your system in practice, doing away with useless places, and making those that are wanted impregnable; moreover, the Artillery and Engineers must be amalgamated and brought by you under one course of instruction. Such, M. le Marquis, is the object of the motion I wish to make. A comparison of the cost of the two systems, the need of establishing a close and detailed scrutiny, the sacred duty of bringing your abilities to deal with the brigandage and charlatanism of the rival corps, no purely military details, or very few, the views of a statesman and a financier; this is what I ask of you as soon as possible, for your own glory and the public welfare: only too happy if I can contribute in the least to put you in your proper place." (ix.)

This might well raise high hopes in Montalembert's breast. Visions of great works opened before him. Two months later we find Carnot writing to him:—

"I accept, general, most gratefully the offer you make to propose me for the construction of your permanent lines, provided that it is, as you say, as chief, and not under the orders of a superior who would perhaps mutilate the projects I might have concerted with you." (x.)

But it came to nothing. The military committee was formed entirely of members of the National Assembly. Other matters occupied Mirabeau's attention. Du Portail, an officer of Engineers, became Minister of War next year, and the reorganization of the corps was carried out under him in a way that was not at all to Montalembert's mind.

But the prospect had its effect on his tone. His eighth volume, published in 1790, is chiefly occupied with a discussion upon the casemated sea-forts newly constructed at Cherbourg, which he declared had been borrowed from him, and marred in the execution by deviation from his designs. The responsible engineer, De Caux, denied his right to claim casemates as his own property, and declared that the projects of the forts were of older date than Montalembert's work. The details had been designed by Meunier, the distinguished engineer who was afterwards killed in the defence of Mayence; and though Montalembert found fault with the gunports, he adopted the principle of them henceforward. It is small blame to him that there was plenty of room for improvement on his designs as regards technical construction, and he may well be excused for some soreness at the persistent withholding of any acknowledgment of his merits; but his absolute and dictatorial tone could not fail to provoke antagonism, and he begins to speak as the jealous patentee rather than as the man making war for an idea.

He appealed to the whole corps of Engineers to judge between him and De Caux, since "the time is past when a few chiefs, by help of an oppressive régime, could coerce opinions and compel silence." The reply to his appeal bore the signatures of several leading officers, but was in fact written by D'Arçon. It sided on all points with the Cherbourg engineer, and repeated many of the old objections to Montalembert's systems.

Turning from that corps in disgust, he then addressed himself to the Artillery :—

"The defence of places depends principally on the effect of artillery, what a disastrous anomaly, then, that it is not left to those who have to produce these effects, to dispose their ramparts in the most suitable way! It is for them, not for the gentlemen of the Engineers, to decide whether these systems would be more easily taken than the modern bastioned systems, and which of the two they would prefer to defend." So also it is for the navy to decide whether his fort would not give them more trouble than the one actually built at Cherbourg.

Carnot wrote to him shortly afterwards (May 1st, 1791), giving excellent advice :—

"You have had experience that it is easier to reconstruct entirely the constitution of a kingdom, than to change the *esprit de corps* of a handful of individuals. . . . However the truth was so overwhelming that you could not help gaining the substance of your cause; you have extracted from your enemies the admission that casemates are excellent, and that if they are not made it is only for want of money. . . One can see how it is; they are unwilling to owe anything to you, but insensibly all our fortifications will be casemated, and Cormontaingne's famous system will be treated as a schoolboy's performance. Like a virtuous man, general, be satisfied with this glory during your life, and look to posterity to render you the justice that is due to you." (ix.)

It might have been thought that other considerations would have forced him to accept this course. He was now nearly eighty years old. He had spent much of his wealth in his various enterprises and in the publication of his works, and the Revolution had robbed him of more. His ironworks had been seized; he had been forced to sell his estate of Maumont for assignats; even his fine collection of fortification models, for which the Empress Catherine had offered him 100,000 crowns, was confiscated. He was reduced to poverty, though he still found money for a draughtsman and a modeller. Moreover, he was not merely an aristocrat, he had been closely identified with the Court. He had held a commission in the light cavalry of the Guard. In 1770 he had married Marie de Comarieu, a lady noted for her talents and accomplishments, who had theatrical performances in which some of the princes took part, and for which her husband wrote pieces that are still to be met with. She was among the emigrants when the Revolution broke out, but he remained behind to carry on the work of his life. At one time he was in imminent danger, and daily looked for his name in the list of the proscribed. But he found friends to shield him. In August, 1793, Carnot became a member of the Committee of Public Safety, and in the following month the Convention accepted from Montalembert a copy of his work, and decreed that something should be done to indemnify and encourage its author. On the 9th August, 1794, he received a formal invitation from the Committee of Public Safety to continue his labours relative to artillery and fortification. A few weeks later, to clear himself perhaps of all Royalist associations, he dissolved his marriage.

"Madame de Montalembert," says Lalande, "remained a little too long in England, and the General was obliged to divorce her. He afterwards married Rosalie Louise Cadet, to whom he had been under obligations at the time of the Terror, and by whom he had a daughter in July, 1796, who was the delight of his latest years."

But this is anticipating. When things were at their worst his pen was still busy, and in 1793 he published his ninth, and as he then thought last, volume. It appeared under a new title—"L'art défensif supérieur à l'offensif." Carnot had found fault with his old title some years before, and Montalembert had defended it on the ground that a title must be short and new, and that it was based upon a feature characteristic and valuable, though not indispensable. But it had become less appropriate as the scope of his work had enlarged, and besides he had found that some officers, confounding perpendicular with vertical, had supposed it to refer to plunging fire. He had long cherished the hope that the French nation, after having had the glory of making the attack superior to the defence, would through him have the still greater glory of making the defence superior to the attack. Looking upon this as achieved, he styled his work accordingly, and had fresh title pages printed for his former volumes.

But in spite of his confidence in his creations, his tone was very bitter and despondent. While declaring that against him and his principles—"The detractors of all useful discoveries have vainly employed their insidious sophisms, conveyed in the pitiable jargon which they make use of when they want to veil their incapacity from the eyes of the multitude;" he owns that "the obstacles raised against the useful changes which might be made in the methods in use, are too powerful for me any longer to flatter myself that I shall see this great good come about."

The chief object of this volume—besides a re-statement of his grievances against the corps of Engineers, a costly and useless body which he would like to see broken up— is to exhibit a further development of his ideas. He was no longer satisfied to obtain fire upon the country from his upper tier of casemates, while the other tiers were hidden. All should see and be seen alike, turning their overwhelming fire to full account, and saving the expense of the heavy excavations and fillings required for his earlier ditches and counterguards. So he proposed to revert to the old style of fortification before the days of cannon. This, his latest system, consists of an angular enceinte with a more commanding circular enceinte within it, both of them wholly of masonry.

There is a covered-way and glacis, and to guard against mining, a deep cunette is carried down to the water level and flanked by a single gun. It was the logical outcome of his principles, but it was also their *reductio ad absurdum*.

This volume soon drew forth a reply, entitled "Des Fortifications et des relations générales de la guerre de Siége," by Michaud, Inspecteur des Fortifications, the citizen style and title of the Chevalier D'Arçon. He, like Montalembert, had run much risk from the storm of the Revolution, and is said to have owed his escape and re-employment to Carnot. It is strange to see those two old officers, whose heads might so easily have fallen side by side under the guillotine, still carrying on this battle of systems with unabated energy. And it is curious also to find that the Prussian Engineers, who, in the next generation drew so much from Montalembert, appreciated no less highly the author of the "Considérations militaires et politiques."

It may not be amiss to borrow another of General Filley's graphic sketches, to convey an idea of the remarkable man who is perhaps chiefly associated in most minds with the unlucky floating batteries of the siege of Gibraltar.

"Excellent sujet," he writes of him during the Seven Years' War, "recommendable par son activité et ses talents, surtout celui de bien lever à la boussole et de très bien dessiner. Il est fort appliqué, a beaucoup d'intelligence, et une ardeur infatigable à la guerre... personne n'aime aussi singulièrement le plaisir et le travail, ne trouve comme lui les moyens et les temps d'y satisfaire pour ainsi dire en même temps. Personne aussi n'est si entier dans son opinion, ne la soutient avec autant de chaleur et de véhémence, et ne trouve tant de raisons pour l'appuyer. Il a foncièrement toute la présomption d'un homme qui a beaucoup lu et médité ; il a de la probité, des mœurs et de la delicatesse dans les sentiments ; en tout c'est un très bon officer, à ne pas laisser en arrière." [9]

Montalembert had had to do with him several years before (in 1774) in reference to some projects for Mauritius. He described him at that time as—

"An officer full of fire, firm and unshakeable in the bastion faith which he professes in all its strictness. He knows nothing of any other methods; he affects to believe it impossible that any other should exist." (ix. 165).

In that case he maintained "in a way so unsuitable and so unbecoming," that engineers alone ought to deal with such questions, that Montalembert took occasion to vindicate (as he was ever ready to do) the right and capacity of experienced officers to design works for engineers to carry out.

In D'Arçon's hands the controversy was carried on with undiminished sharpness and by no means perfect taste, but with far more skill and judgment.

"The officers of Engineers will admit," he says, "that their art has not reached perfection; that led away by the authority of Vauban's genius they have perhaps endeavoured to imitate him too servilely, or have not been always happy in their application of his principles."
... "We will own further that some old Engineers have been more than cold as regards the very great advantages of good casemates; they have made the

[9] Augoyat, "Spectateur Militaire," 1836.

mistake of supposing that the inconvenience of the smoke was very difficult to remedy. But with the same frankness we assert that we have not been sufficiently peremptory in our rejection of those proposed by Montalembert."

It is plain that the casemate question has run its three stages: it is not true; it is contrary to orthodoxy; every one knew it before (except some old engineers who are thrown overboard). He makes merry with Montalembert's last compositions, and with the costliness of his constructions, especially if applied on the gigantic scale he desires, for the fortification of the frontier. It would be necessary to sell France three times over to raise the money, and as for the men, there would be four pieces of cannon per head with the army on a peace footing, "and in extreme crises every soldier besides his musket would have his big gun to work."

But there is more interest in following his method of attack, for he does not ignore their powerful armament. Indeed a saying of his, with reference to the siege of Gibraltar, had been quoted approvingly by Montalembert, that—

"We can count upon nothing in war if we may not take it as a principle that six pieces of artillery will in all cases silence one piece."

However, he undertakes to overpower these impregnable works with forty heavy guns in five days. He will open a sort of parallel a quarter of a mile long at about 400 yards from the walls, having a parapet twenty feet thick. The forty guns will be placed at intervals along this parallel-battery and mounted on elevating carriages, so that—

"With a turn of the hand the pieces will be raised to fire *en barbette* against the walls of the casemates : after the shot, the mere recoil of the piece will bring it at once under shelter of the epaulment by sinking it twenty-two inches."

Or if the efficacy of such a carriage, as yet untested, should be disputed, there is another method. The epaulment above mentioned may be used merely as a screen, being made seven or eight feet high, or even more if necessary. A second epaulment for the guns will be made behind it and parallel to it, the guns being thirty feet apart. The embrasures in this epaulment will be as small as possible and framed with solid timbers. Portions of the crest of the front epaulment will then be cut away so as just to allow the guns to see the highest tier of casemates at certain points agreed upon. Thus bit by bit this highest tier will be silenced and breached, bringing down with it the arch which covers the whole structure ; and the lower tiers can be dealt with afterwards.

As he justly says :—"There is nothing that men can build which men in larger numbers will not be able to destroy," if they are left long enough without interruption ; and places can only become impregnable by the co-operation of troops outside with energetic garrisons, the fortresses giving time for the formation of field armies, and serving as pivots for them, and the field armies then delivering the fortresses. While vindicating the bastioned trace in the abstract, he allows that it is often unsuited to ground, and that quite recently the French engineers had found occasion to recommend masonry block-houses for closing defiles on the frontier.

"We would willingly leave Montalembert the satisfaction of claiming these as his invention, if only he will be kind enough to recognize the difference of application."

Is it impossible, he goes on to ask, to arrive at some compromise, "to draw from the leisure work of an irreconcilable amateur some useful ideas," and turn casemates to account for defence as well as for shelter?

"An indispensable preliminary will be to hide them from the batteries of the attack, so as to save them from inevitable and prompt destruction. But how cover buildings from fifty to sixty feet high? By reducing them from five or six stories to three only, including the ground floor and the platform, which can be brought within a height of twenty-two feet."

This can easily be covered by an earthen rampart, the ditch of which must be thoroughly flanked, either by means of a bastioned trace, or where that is unsuitable, by reverse fire from casemates under the counterscarp at the salients. What is here sketched had already been carried out by him in the type of work known as D'Arçon's lunette, one example of which lately played a part in the defence of Strassburg. Montalembert had found great fault with that type already, but it is to his partisans rather than to himself that D'Arçon appeals:—

"Among the disciples seduced by a composer as adroit as he is indefatigable, there are, we doubt not, judicious men capable of listening to reason."

Carnot, it must be remembered, was now controlling the War Department. But Carnot, as one finds in the "Défense des Places Fortes," was by no means a thorough-going believer. He considered that casemates which could be seen from the country would infallibly be destroyed, and he had no wish to let Montalembert, especially in his latest mood, rule despotically over the theory and practice of fortification. The two last volumes of Montalembert's work—the tenth published in 1795, and the eleventh about three years later—are collections of miscellaneous occasional papers, chiefly giving vent to his vexation and disappointment at the continued vitality of the doctrines of the old Engineer school of Mézières. One of these pieces was provoked by the publication of Noizet de St. Paul's "Traité complet de la Fortification;" others by the "Journal de l'École Polytechnique." He was shocked to find that in this new school, started under the auspices of Carnot and Prieur, the teaching of fortification was proceeding upon the old lines, and the writings of his adversary, D'Arçon, were held in high esteem. As he complained to Bossut, one of the examiners—

"The engineer Say teaches at that school the art of defilade without mentioning my opinion as to the uselessness of that method."

He had also a battle to fight with some leading officers of Artillery on behalf of his gun-carriages. But he had at all events one most ardent Artillery disciple in General Bélair, the author of "Nouveaux Élémens de Fortification." This officer undertook a military dictionary, which was intended to make amends for the complete ignoring of Montalembert in the military section of Diderot's "Encyclopédie," and overflows with fulsome laudation of him. The first portion of it is printed in his tenth volume, but fortunately the compiler was appointed to a command before the letter A was finished.

We may leave Lalande to describe the last two years of Montalembert's life.

"He consoled himself with incessant work. I have many interesting memoirs by him on different objects affecting the public welfare, on politics and administration, which are the fruit of ripe reflection, and of great experience, for instance, on the descent on England projected two years ago (1798); he added verses which ended—'L'on ne vaincra jamais les Anglais que dans Londres.' His activity had not suffered from his great age. Only a few months ago he read to the Institute a new memoir on naval carriages; he was received with veneration and listened to with a religious silence, one had never heard a man of eighty-six read with so strong a voice a memoir of so much importance to France. The Institute wrote to the Minister of Marine, who sent orders to Brest to have the carriage made.

"He was on the list of candidates for a seat in the Institute, and his name had in fact been put first by the mechanical section; but hearing that General Bonaparte was talked of as a candidate, he wrote a letter expressing his wish to see the young conqueror of Italy receive this additional honour.

"His mental power continued to the end; a month ago he wrote some observations on the siege of St. Jean d'Acre, which went to corroborate his system of defence. But the winter, which has produced so many illnesses this year, gave him a cold which turned to dropsy, and this carried him off on the morning of the 7th Germinal."[1]

What, it may be asked, was the net result of his forty years of strenuous labour upon the art of fortification? It

[1] 27th March, 1800.

is possible to argue, as Fourcroy did, that what was true in his work was not new, and what was new was not true Absolute originality is hardly ever combined with absolute freedom from mistakes. Much of what he said had been said before, by Rimpler, Landsberg, and others, but it was he who made the world listen to it. He lifted the art out of a rut in which it threatened to stick fast, and gave it an impetus it has never lost. His projects were not fit to be carried out as they stood, and they have become much less fit with the lapse of a century. But they have furnished numberless hints to other designers, from D'Arçon downwards. Detached walls and casemated keeps and caponiers have since been the stock furniture of the military architect. His principle of energetic artillery defence from the outset has now become an axiom; and though his provision for that purpose, his unmasked upper batteries, have been universally condemned, yet the various devices which have been brought forward since,—Haxo casemates, mortar casemates, blinded batteries, and turrets,—show at once the importance and the difficulty of the problem of furnishing protected fire upon the country, which he too lightly thought he had solved.

CARNOT

CHAPTER IV.

CARNOT.

AMONG the leading actors of the first French Revolution, there are few figures more remarkable than that of Carnot, whether we look to the varied incidents of his career or to his own character and capacity. Captain of Engineers, Republican commissary, colleague of Robespierre, on the Committee of Public Safety, Member of the Directory, Minister of Bonaparte, and Governor of the fortress of Antwerp; between whiles a laborious *savant*, living in exile or in retirement; he comes to the front again and again at each turn of the national fortunes.

And yet this was far from being due to that adroit suppleness which adapts itself with equal ease to every form of government. On the contrary, Lord Brougham wrote in 1814,[1] "It is undeniable that he has shown himself the most inflexible friend of liberty whom France has produced—the man most renowned for acts of personal opposition to tyranny of every description."

"All acknowledge," to quote further from the same source, "his vast political and military talents, crowned upon every occasion with extraordinary success; his genius

---

[1] *Edinburgh Review*, Vol. XXIV.

for the abstract sciences and his contributions to their progress, unequalled by those of any other man not a mere philosopher. Upon these points there can exist no difference of opinion; but it is singular to find an equal unanimity in extolling his integrity as a public man, notwithstanding the horrid scenes of faction in which he has borne sway, and the manifold contaminations by which his course has so constantly been surrounded."

An excellent sketch of his career has been given by M. Arago in his "Scientific Biographies," in which full justice has been done to his work in science; we propose to dwell here more especially upon his military writings, illustrated as they are by the copious "Mémoires sur Carnot," published by his son in 1861.

Lazare Nicolas Marguerite Carnot was born at Nolay, in Burgundy, on the 13th of May, 1753, and was the second son of Claude Carnot, a notary. The bent of the future "Organizer of victory" showed itself early. When he was ten years old he was taken to a play at Dijon, in which the attack of a fortress formed part of the piece. After watching it with deep attention for some time he suddenly rose, and in spite of all efforts to stop him, abruptly addressed the general on the stage. He pointed out to him that his artillery was badly placed and his gunners exposed to fire from the fortress, and urged him to place a battery behind a certain rock which would give better shelter.

He was educated at the College of Autun, along with Joseph and Lucien Bonaparte. At the age of sixteen he was sent to Paris to study mathematics, his marked turn for the exact sciences having decided his father to train him for the Engineers or Artillery. At Paris he attracted the notice of D'Alembert, who predicted great success for him, and exercised much influence on his scientific development. He was less fortunate with Rousseau, to whose house he made a pilgrimage with one of his school-fellows. The young enthusiasts met with an ungracious reception, which discouraged further visits of the kind. The religious principles in which he had been brought up at Nolay and Autun, and which had taken strong hold of him, met with many shocks in the atmosphere of Paris. Impatient of doubt, he set himself to study theology for eighteen months; and ultimately fell back, like most of his contemporaries, on simple deism.

In 1771 he passed third for the corps of military Engineers, and entered the School of Mézières as second lieutenant. There he became a close ally of Gaspard Monge, the inventor of descriptive geometry, who was assistant-professor of mathematics, and under his guidance Carnot made rapid progress in mathematics and physical science.

At the beginning of 1773 he left the school and went as first lieutenant to Calais. He spent the next eighteen years in this and other garrisons in the north of France, becoming captain by seniority in 1783. Up to the time of the Revolution," he says, "I lived from day to day like so many other young men, especially in the army, passing from one station to another without even noting the dates." His mind, however, was by no means inactive during this time, and he was already looked upon as " an original." While writing songs good enough to take the fancy of young Béranger, and while fellow-member with Robespierre of the Literary Academy of Dijon, he also studied and wrote on scientific and military subjects.

In his "Essay on Machines," published in 1783, he developed what has since been know as Carnot's theorem, respecting the loss of force consequent upon all abrupt changes of velocity. In the following year he sent to the Académie des Sciences a detailed memoir on balloons, the new wonder of that day, which he proposed to steer by means of an apparatus of wheels, of which heat was to be the motive power.[2]

The Academy of Dijon, which might claim to have discovered Jean-Jacques Rousseau by its competition of 1750 (on the influence of the arts and sciences), offered in 1784, a double medal for an *éloge* of the great Burgundian, Vauban; Carnot, a son of the same province, obtained the prize, and soon afterwards pawned the two medals to assist a brother-officer in distress.

Carnot's *éloge* is divided into two parts: he first follows Vauban's military career, and then treats of him as a citizen and a statesman. He does full justice to Vauban's great merits in these latter relations, and dwells with sympathy upon his care for the working classes and his efforts to lessen their burdens and improve their condition. "What," he asks, "should be the object of government but to compel all the members of the State to work? and how bring them to this but by transferring wealth from the hands where it is superfluous to those where it is wanted? providing the one with the means of industry, and depriving the other of the means of idleness. The idler is of no use until the moment of his death; he enriches the soil only by his own return to it. And yet it is

---

[2] Ten years afterwards, when controlling military affairs, he gave balloons their first trial in military reconnaissance at the battle of Fleurus, where a captive balloon proved very useful.

this idler who enjoys everything, while the wretched cultivator has toil and humiliation for his cup; the trunk dries up, while the useless and parasitic branch devours the whole substance."

But there is not so much discrimination as might have been looked for in Carnot's sketch of the military engineer. Montalembert published anonymously an edition of the *éloge* with annotations of his own, and his caustic remarks upon the inflated language and the looseness of the statements are usually well founded. Vauban's improvements on the art of attack are, as Montalembert justly says, his true titles to distinction. "If to these be added the merit of sixty years of the most assiduous and perilous service: the merit of the order and economy which he contrived to establish in all the works that he directed (an advantage all the greater because it has survived him), the merit of his patriotic zeal and attachment to his King, lastly, the merit of all the good he did and all that his writings show he would have liked to do; under these several aspects combined, there is no man who should seem greater or should attract more homage, and no one whose memory should be held more dear."

But it is too much, even in a panegyric, to speak of him as a creative genius in fortification, with boldness to overturn and wisdom to reconstruct, who has got rid of the false maxims of his art, and who has left nothing for his successors to do but to imitate him. In the 160 places that he helped to fortify, abundant as are the marks of his skill and judgment, there is scarcely any mark of his originality, except in the tower bastions of Belfort, Landau, and Neuf-Brisach, which were, it is true, his chief pride, and had great merits, but which were never adopted by other engineers either in France or elsewhere. Carnot himself, in fact, declared later that "this illustrious engineer, always occupied with the attack, did nothing of importance for the defence."[3]

It was partly this misplaced praise of Vauban as a fortress-builder, that led Montalembert to take up and annotate Carnot's performance; but partly, also, certain disparaging phrases which he assumed to be directed against himself. He was already smarting under the "Mémoires sur la Fortification Perpendiculaire, par plusieurs officiers du génie," as mentioned in the last chapter, which showed the hostility felt towards him by the heads of the corps; and in his answer to the "Mémoires," he again noticed Carnot's supposed allusions to him. It is characteristic of Carnot that he did not allow himself to be drawn by desire of advancement, or driven by mortified vanity, into the fashionable attitude of contempt towards the innovating cavalry officer, or even stand neutral in the controversy. He wrote at once to Montalembert, assuring him that far from having alluded to him in the *éloge*, he was at the time quite unacquainted with the "Fortification Perpendiculaire." "I have read it since," he adds, "this work so admirable, so full of genius; I have done justice to it on every opportunity that has offered, and I shall not cease to do so, having nothing so much at heart as to know the truth, and to see light shine, come whence it may. Now that your casemates are known and have been tested, fortification will wear a new aspect and become a new art. It will no longer be

---

[3] "Traité de la Défense des Places Fortes." Discours préliminaire.

allowable to fortify without them, nor to employ the public money in mediocre works, when one has the means of making good ones. The whole credit will be yours; no one can dispute it with you. It is certain that you have solved the problem long studied in vain, how to obtain well-protected fire readily and in abundance, the most important problem of military architecture."

At the same time he begs Montalembert not to be too impatient for the general acceptance of principles so revolutionary. "You yourself would look more closely into them, and maybe would alter many things in your projects, if it were a question of carrying them out." He concludes by denying that the corps of Engineers, as a body, is averse to Montalembert. Although not borne on the list of the corps, "we none the less consider that we have a right to reckon you among its most distinguished members."

This letter was published by Montalembert. It was not likely to commend the writer to his superiors in the corps; and a short imprisonment which Carnot afterwards underwent in the Castle of Béthune, was ascribed by him to the disfavour into which it brought him. The ostensible ground, however, for this imprisonment, was his irregular conduct in quitting his station without leave to fight a duel at Dijon with a man who had supplanted him in a love affair.

In the following year (1789), he addressed to the National Assembly a formal "complaint against the oppressive *régime* by which the corps of Engineers is governed," in which he argued for freedom of discussion on professional questions, and took occasion to dwell on the advantages of Montalembert's casemated batteries.

The creation of the frontier lines of fortresses, the combination and connection of individual places in one general system of national defence, was a side of Vauban's work which had not been overlooked by Carnot in his *éloge*. He had noticed the several functions of fortified towns, "now serving as a retreat for a beaten army, gathering up its fragments, restoring order to it, and reviving its courage; now as a bulwark, barring the way to an enemy in country which nature has left without defence; often as an entrepôt securing the subsistence of an army which protects the frontier or carries the war beyond it; sometimes as a link to keep up needful communication between different provinces." And he had also touched upon the further question :—"But is there no danger in an excessive number of fortresses? Is there not a limit at which the general policy requires that we should stop? Doubtless there is, and it is this limit that Vauban seeks to discover; he is not unaware that one can supply the place of fortresses by men, that one may defend with human arms instead of walls; he recognizes this fundamental truth, that from whatever point of view one regards fortresses, their sole object, when we come to the root of the matter, is to save expenditure of men, that wherever they do not fulfil this object, they are superfluous; that they become injurious to the state when by their multitude they have an opposite effect; and that this luminous maxim should exclusively regulate their number and their disposition."

In 1788, a "Recueil de Mémoires sur la trop grande quantité de Places de guerre qui subsistent en France," was officially published, chiefly through the influence of Guibert. This led Carnot to return to the subject, and he addressed a forcible memoir to the Council of War in defence of the system of frontier fortresses.

He first sets himself to show that "the interest of a state, such as France, and situated as she is, is to renounce altogether the spirit of conquest, and to confine herself to the defence of her possessions," and then that the best means thereto is to have many and strong fortresses. He appeals to history to show the disasters brought upon the country by wars of conquest, such as those of Louis XIV., and the value of her fortresses in adverse times.

With the same faith in paradoxical arithmetic which reappears later, he argues that since the proportion of besieged to besieger may be as one to ten, to pass from the defensive to the offensive both men and material must be multiplied one hundred-fold, "and that on the other hand, he who passes from the offensive to the defensive saves, simply by this change of system, ninety-nine hundredths of his resources of all kinds."

He goes on to examine the objections brought against fortresses, and more especially the objection that the money spent on them would be better spent in strengthening the active army. His answer to this is, that on the contrary, they afford the means of maintaining a stronger army than would otherwise be possible. In the absence of fortresses, the whole army must be always on foot ready for sudden emergencies; but with a well-fortified frontier, it will ordinarily be quite sufficient to keep under arms say 100,000, in place of 200,000 men that would otherwise be needed, and the economy resulting from this will admit of the army being raised to 300,000 when the crisis arrives.

Fortresses, it is said, serve only to ward off blows: they cannot help you to put pressure on rival powers, or to avenge their insults. But on the contrary, it is the nation that is without fortresses that is most helpless in these respects. If it attempts to carry the war into the enemy's country, its own territory stands exposed to his diversions, and if its troops are defeated, there are no bounds to its disasters.

But places, some contend, may be improvised when and where they are wanted; why go to the cost of building and maintaining them for perhaps a century before they are attacked? Because field-works, hastily thrown up, can ill supply the place of regular fortresses, and they would not in any case fulfil the main object of the latter, to secure the kingdom against an unexpected attack. That fortresses are seldom besieged is no proof of their inutility: "A place that one should succeed in making absolutely impregnable, and that should be known to be so, would assuredly never be besieged, and yet none could be so useful, for it would stop the enemy short, and a cordon of such fortresses would assure us eternal peace."

The value of fortresses would not be questioned, it is said, if they were impregnable; but at best they only delay invasions, and always yield in the end. This is a reason for making them stronger, not for doing away with them, and there is no limit to the strength they may receive Besides, "granted that our places are not impregnable, but are our armies invincible? What right have troops who let themselves be beaten to require that towns should not let themselves be taken? . . . Suppose any of our places could only hold out for six weeks: what matter, if six weeks is enough for succour to arrive."

But it is complained that these artificial defences swallow up all the treasure of the state. The answer is that they cost less than one-thirtieth of the expenditure upon the regular army. "Let it be shown that those troops render thirty times the service of the places, or even that they render as much service."

There still remains, however, the weightiest objection: "Nobody, it is said, wishes to decry the utility of fortresses, but there are too many of them." Carnot's reply is to ask whether new and stronger ones are not needed, since none of those existing could hold out for three months, and to ask also why people should fear an excess of these any more than of nature's fortifications—defiles, rivers, and forests. But his study of Vauban shows that he was not really blind to the force of these objections, which the next twenty years did so much to illustrate. In 1812 he wrote: "The opinion of the greatest generals has always been that it was much better to have a few fortresses of large size, in good condition, and well supplied with everything necessary for their defence, than a large number of indifferent or neglected fortresses; for small scattered garrisons cannot stop a hostile army, which usually contents itself with watching them by detached corps while it moves on, until want compels them to surrender one after the other. . . . . The suppression of small places that have no definite object, far from being an inconvenience, would be of advantage to the defence of the frontiers, if the sums spent yearly on their maintenance were diverted to fortresses of the first class."[4]

---

[4] "Traité de la Défense des Places Fortes," p. 506.

He follows up the arguments of his Memoir by two proposals :—first, to repair the existing fortresses and to supplement them by ten grand places absolutely impregnable, either new or converted; secondly, to substitute for the 170,000 regular troops and for the militia, a force of 300,000 men thoroughly disciplined, of whom one-third only should be kept with the colours in time of peace.

This body of 100,000 men is like a "kind of military school where each of the 300,000 soldiers of the Army should go in his turn; and when, at the end of a year or eighteen months, more or less, he should be sufficiently trained, he would be sent home to finish the term of his engagement (from four to ten years) on condition of presenting himself again when wanted."

In case it should be thought that this is much too short a time to make a soldier, he reminds his readers that "the engagement of the Swiss in the service of France is only for four years, and every one knows with what precision they manœuvre. The national troops of the King of Prussia are assembled only for six weeks in the year. We imitate him in everything except in that which is best about him and most agreeable to the genius of our nation. No doubt veterans are very valuable, but a veteran is a man who has seen service. A man who has done nothing but strut about a barrack-square for eight years is just as raw a soldier as the man who has done the same for six weeks." We seem to be listening here to a writer of the nineteenth, rather than of the eighteenth century.

The reform of abuses and the attacks on the privileged orders which marked the outbreak of the French

Revolution, found in Carnot a ready and outspoken sympathizer. He became president of the popular society of his garrison town—Béthune; and in 1791, immediately after his marriage, he was elected deputy for the Pas-de-Calais to the Legislative Assembly. His younger brother, known as Carnot-Feulins, who was also an officer of Engineers, was elected a deputy at the same time.

In the Assembly he was made a member of the Military Committee and at once took an active part in military questions, but rather as a democrat than as a soldier. He proposed that the townward fronts of all the citadels should be levelled. "A citadel," he said, "is a monstrosity in a free country, a harbour of tyranny against which the indignation of the people should rise." The question of military obedience was raised in the Assembly, owing to some orders issued by the Minister of War. Carnot argued that passive obedience should be exacted only in presence of the enemy; that the soldier when employed at home on police duties, becomes a national guard, and ought only to be subject to the common laws and to yield an *obéissance raisonnée*.

On the 24th of July, 1792, he brought forward and carried a proposal for the manufacture of pikes to arm volunteers, as there were not enough muskets and there was no time to make them. He supported this proposal by arguments and instances in favour of the pike as a weapon. Eight years before, in his *éloge* of Vauban, he had reserved his opinion as to the merits of the change made by him in substituting the bayonet for the pike.

But strongly democratic as he was, Carnot never

pandered to the passions of the clubs. He reported in
the severest terms on the troops who had murdered
General Dillon at Lille; and when he was sent to
inquire into the rumoured royalist poisoning of
volunteers at Soissons, by mixing powdered glass with
their flour, he showed, to the great vexation of the
patriots, that it was a mere accident, due to a broken
window.

He was one of the Commissaries sent by the Assembly
to the Army of the Rhine immediately after the suspension of the king, to explain the situation and to deprive
of their commands all officers who refused to recognize
the new government.

At the end of August he was again returned by the
Pas-de-Calais to the new Convention, and in the following month he was sent with five other deputies to
organize the defence of the Pyrenean frontier. He
returned in the earlier part of January, when the trial of
Louis XVI. was in progress, and voted for his execution
on grounds of "justice and policy alike."

Soon afterwards Carnot was sent to the north, and
in the beginning of April he was appointed one of the
five deputies charged to bring Dumouriez to the bar of
the Convention. His nomination fortunately reached
him too late, otherwise he would have shared the fate
of his colleagues who, together with the Minister of
War, De Beurnonville, were handed over by Dumouriez to the enemy, and remained prisoners for some
years.

On the 14th August, 1793, while still in the north, he
was chosen a member of the Committee of Public Safety
—that "'Committee of Public Salvation' whereat the

world still shrieks and shudders."[5] This Committee had been formed in April, "to stimulate and watch the ministers, and, when necessary, to suspend their orders."[6] Composed at first chiefly of Dantonists, it came by its reconstitution in July more under the control of Robespierre, who himself entered it at the end of that month. The want of military experience led to the enlargement of the Committee from nine members to twelve. Two engineer officers, Carnot and Prieur (de la Côte d'Or) were brought into it, the former taking charge of the *personnel* and movements of the troops, and the latter of the manufacture of arms and other accessory services.

Robespierre opposed Carnot's election, and was always hostile to him. "We need your services," he once said to him, "and we therefore tolerate you in the Committee; but remember at the first disaster of the armies you will lose your head." "Oh that I could get to understand these cursed military matters," he cried at another time, "and be able to dispense with that intolerable Carnot." The aversion was, of course, heartily reciprocated.

The task imposed upon Carnot on entering the Committee was formidable enough. The Minister of War, whom he was to stimulate and watch, was Colonel Bouchotte, an ally of Marat and the Hotel de Ville, whose ministry stands unrivalled in the number and flagrancy of its scandals. He caused the lampoons of Hébert to be circulated by millions in the camps, always took the part of mutinous soldiers against their officers, and maintained the most worthless and incapable men in command.[7]

[5] Carlyle.   [6] Sybel.
[7] On the 1st April, 1794, the Ministry of War, together with the

At the beginning of 1793 the number of troops present under arms throughout France was less than 200,000 men. Already at war with Austria, Prussia, and Sardinia, the Republic was on the point of war with England, Holland, and Spain. The regular troops had lost both numbers and *morale* by the events of the Revolution. The bonds of military discipline had been broken, and distrust and dissension prevailed on all sides. Revolutionists excited the men against their officers, of whom at least two thousand had resigned by the end of 1791, and many more followed when the king was set aside in 1792. The officers of the several arms were at that time the chief agents in recruiting, and as neither those who left nor those who remained cared to be active in the matter, it ceased almost entirely, and vacancies in the ranks were unfilled. This was all the more the case because the volunteers enlisted under the Decree of the 4th June, 1791, being more highly paid than the regular troops, not only drew away the recruits who would otherwise have joined the latter, but also stimulated desertion from their ranks.

At the same time these volunteer battalions, in which the officers were elected by their men, and promoted according to length of service, were undisciplined and of little value.

The unlooked-for successes of 1792, which, according to Marshal Bugeaud, were due to the regular troops, were followed by reverses in the early part of 1793. By

other five ministries, was suppressed, at Carnot's instance. Its duties were transferred to a Commissioner of the organization and movement of the land forces, under the immediate control of the Committee.

Dumouriez's defection, the Republic lost its ablest general, and others were appointed, removed, and guillotined, in quick succession. The need of much larger forces was urgent, but the distrust of the democratic leaders towards the regular troops remained undiminished. On the 24th February, 1793, a levy of 300,000 men was decreed, to be raised either by voluntary enlistment or conscription, from men between eighteen and forty. To assist in raising it, Conventional Commissaries were, upon Carnot's proposal, spread all over the country. Owing to them the levy was carried out without much difficulty or delay, "and became, owing to the successive reinforcements which it furnished to the French armies, the true cause of the superiority which they soon assumed over those of the Foreign Powers, formidable as they were."[8] By the end of July the numbers present under arms had risen to nearly 500,000 men.

But, for the time, the new levies were powerless to affect the course of the war. In July, Mayence and Valenciennes surrendered, and the road to Paris lay open to the allied armies. The cry was raised on all sides for fresh troops, and on the 23rd August, nine days after Carnot joined the Committee, the Convention decreed a *levée en masse*, or universal conscription of all citizens between eighteen and twenty-five years of age. This raised the strength of the armies to 700,000 by the beginning of 1794.

But the men had not only to be raised, they had to be converted into soldiers, and this was a work of more time and difficulty. Bouchotte, the Minister of War, who

[8] "Tableau Historique de la Guerre de la Révolution de France, 1792-4."

shared the Jacobin hatred of the regular troops, was anxious that the new recruits should not be exposed to the evil influence of close association with them. He ordered, therefore, that the recruits of each district should be formed into a battalion, choosing their own officers, and the battalions formed into a division before marching to the frontier. But these divisions were so ill-trained and ill-disciplined, that a radical reorganization was felt to be necessary.

Already, in February, 1793, in order to "democratize the army," and get rid of its old traditions, a motion had been carried by Dubois-Crancé for forming the whole of the infantry into demi-brigades, each consisting of one line and two volunteer battalions; but it was found impossible to carry out this change when the campaign was just beginning. On the 22nd of November the Convention repeated its decree. The battalions of the new levy were broken up, officers and non-commissioned officers reduced to the ranks, and the whole fused with the older battalions of volunteers. The strength of the demi-brigade was fixed at 3200 men, and 209 demi-brigades were formed. These were re-numbered by lot; all distinctions of pay and privilege were abolished; the white uniform of the regulars was discarded, and the blue uniform of the volunteers was made universal. "Farewell," says the Duc d'Aumale, "to the old names—Picardy, Champagne, fearless Navarre, unstained Auvergne ! But the numbers of the demi-brigades soon had their own aureole of glory. Who would not have been proud of belonging to the 'invincible 32nd,' 'the terrible 57th,' 'the intrepid 106th'?" [9]

[9] "Institutions Militaires de la France."

The demi-brigades were coupled to form brigades; two or three brigades formed a division; and an army consisted of several divisions according to the work assigned to it. Henceforward the component battalions of brigades and divisions were to remain always the same. The battalion served as the tactical, the brigade as the administrative, and the division as the strategical unit.

This was perhaps the most important step that has ever been taken in army organization. Up to that time, regiments though temporarily associated together in first line, second line, or reserve, had no real bond of union, either with each other or with their commander, who was often merely detailed for the day. Other countries soon followed the example of France in adopting the divisional system.

One most necessary point in the reorganization of the army was to get rid of incapable officers; but this was extremely difficult, owing to the protection given by the Jacobin leaders to their own adherents. About four thousand were removed by means of a regulation that all officers must be able to read and write. It had been decreed in February that, up to the rank of brigadier, one-third of the promotions should go by length of service, and two-thirds by election by subordinates. As regards the former, length of service in the grade was afterwards substituted for length of service in the army; and election, while maintained in theory, was practically confined to the lower ranks.

The staff of the armies was carefully chosen, and attached, not to the general, but to the force; so that a change of commander, only too frequent at that time, might cause as little confusion as possible.

The Polytechnic school was founded, and the special schools of application were revived, to give instruction to the officers of the scientific corps. The Engineers had been reduced to a corps of officers only. Carnot restored to them their companies of miners, which had been attached to the Artillery, and added twelve new battalions of sappers. "Enrolments by bounty and arbitrary recruiting abolished, the obligation of military service imposed on all and accepted without resistance, the unity of the army restored and marked with the national stamp, the mode of promotion fixed by law, scientific and military education secured to the officers of the special arms, the duties of the generals defined, the principles which should govern the formation of the active armies laid down and put in practice, the Roman legion revived in the French division, such are the steps made under Carnot's administration. All was not exclusively his work, but he had the merit of causing what had been tried with success here and there, to be carried out everywhere, and extending to all the benefit of the experience so promptly and so dearly acquired by some few."[1]

But he did not confine himself to military organization; he was equally active in directing the operations of the armies. It was under his instructions for the relief of Dunkirk, that Houchard won his victory of Hondschotten, in September, 1793; and when Houchard had been replaced by Jourdan, and sent to the guillotine for not improving his success, Carnot did not shrink from giving the new commander the support of his presence, and accepting at Wattignies the full responsibility for the movement

[1] Duc d'Aumale, "Institutions Militaires de la France."

that decided the fate of the day. Napoleon spoke of this battle afterwards as "le plus beau fait d'armes de la Révolution," and gave all the credit of it to Carnot. Jomini has mentioned it as an instance of his dangerous system, alike in tactics and in strategy, of operating simultaneously on both wings. But this system was connected with general conceptions which lay at the foundation of the Revolutionary method of war. "Carnot," said Dumouriez, "is the creator of the new art of war in France, which Dumouriez had only time to indicate, and which Bonaparte has brought to perfection." He saw that to beat the better-disciplined troops opposed to them, and to turn their own enthusiasm to full account, the French must fight in a new fashion, and must always be the assailants. This demanded local superiority of numbers, notwithstanding general inferiority; a thing possible only by perfecting the transport service so as to secure mobility, and by maintaining the frontier fortresses in an efficient state. By these means, full advantage might be taken of the unity of direction on the French side, and the divisions and jealousies on the side of the Allies.

Above all, Carnot exerted himself to raise the *morale* of the troops, and to combine in them patriotic ardour with soldierly discipline. He sent them fervid commissaries to stimulate them, but he took care to give them leaders of military capacity, and this with little regard to party complexion. He ventured even to protect and employ some of the most eminent men of his own corps— such as D'Arçon, Marescot, and D'Obenheim, who stood in danger as aristocrats or royalists; and he obtained a decree from the Convention enjoining Montalembert— Marquis as he was—to continue his labours on fortification and artillery undisturbed. He early recognized the ability of Hoche. When the latter, at that time unknown, sent to the Committee a memoir on the conduct of the war in Belgium, the comment of Carnot was—" this is a man who will make his way;" the comment of Robespierre, always in dread of a military dictator,—" this is an exceedingly dangerous man." And when Hoche was afterwards removed from the command of the Moselle army by St. Just, he was saved by Carnot from the Revolutionary Tribunal. On the other hand, much against the wishes of Robespierre, Carnot procured the recall of the brutal and incapable Turreau—the organizer of the "infernal columns"—from La Vendée, and introduced the less savage system of warfare which gradually led to its pacification.

The fruit of Carnot's work, already visible in 1793, was seen more abundantly in the successes of 1794. Of all the enemies of the Republic none was so much to be feared as England. The others were half-hearted in the war, jealous of each other, and crippled by internal disorder or by want of money. An invasion of England was consequently the scheme which Carnot had most at heart, and as a stepping-stone to that, the occupation of the Netherlands was of great importance. Accordingly it was towards the north that the main strength of the French forces was directed. The victories of Tourcoin and Fleurus drove the Austrians out of Belgium, and before the end of the year Pichegru had conquered Holland, and the French Republic, no longer intent upon mere self-defence, was claiming cessions of territory as a condition of peace.

Meanwhile much had happened in Paris. Robespierre had fallen in July. On his fall the various factions crushed by him began to revive; the Committee of Public Safety lost its authority, and before long demands were made in the Convention for the indictment of Collot d'Herbois, and others of its old members. They were arrested in October, but their trial by the Convention did not come on until the latter end of March, 1795. Carnot had usually acted with them on the Committee, sharing their hostility to Robespierre, and finding them more zealous than others for the energetic prosecution of the war. He, with Prieur and Lindet—all three known as "the workers" of the Committee—had claimed to be included in the arrest of their colleagues, and when the latter were tried, spoke strongly in their defence.

Carnot pointed out that the division of labour which had been imperative among the several members of the Committee, rendered each one morally responsible only for the papers of his own branch, even though as a necessary formality his signature was appended to others. He said that he had often himself so signed without reading. If some of their acts had been ultra-revolutionary, it must be remembered that public opinion had been the same. Men who had toiled so hard for their country, and had delivered her from such straits, ought not to be harshly judged, least of all at a time when clemency was extended even to Vendean rebels. By condemning them, the Convention, which had acquiesced in their acts, and obeyed the same popular impulse, would in fact condemn itself, and it was this result the enemies of the Republic were really so eager for.

Sentence of exile was ultimately passed on the accused, and on the 29th May their accusers proceeded to demand the arrest of other leading actors in the days of the Terror.

Carnot's own name was pronounced. There was an anxious silence, till some one cried out, "Will you dare to lay hands on the man who has organized victory in the French armies ?" The happy phrase, "he has organized victory," was caught up and repeated with enthusiasm, and the accusation was dropped.

In the early part of March, Carnot had definitively left the Committee of Public Safety, and refused to offer himself again for re-election. On the 11th May he became *chef de bataillon* in his corps by seniority. He had made no use of his power for his own advancement in this or any other respect. "After every journey undertaken on the public service, he conscientiously returned the money he had not used to the Treasury, to the great annoyance of the officials, who had no columns for such an entry in their registers."[2]

He did not long remain out of office. In October, 1795, when the executive Directory was appointed, he was chosen a member of it, together with another Engineer officer, Le Tourneur. He was again entrusted with the direction of military affairs, and the supervision of the War Ministry, which had been abolished in the spring of 1794, but was now revived.

He found things in a very different state from that in which he had left them. Owing to Pichegru's treachery, the siege of Mayence had been raised, and the Rhine frontier lay open to the enemy; the war had broken out afresh in La Vendée; and in Italy the Republican army was passive and destitute.

Pichegru was soon replaced by Moreau in the command of the army of the Rhine; Hoche was sent to La Vendée;

[2] Von Sybel.

and Bonaparte was placed at the head of the army of Italy. "It was not Barras who proposed Bonaparte for the command of the army of Italy; it was myself," wrote Carnot afterwards; "but they waited a little while to see how he would succeed, and it was only among his intimate friends that Barras boasted of having been the author of the proposition made to the Directory. If Bonaparte had failed, the blame would have been laid on my shoulders; I had chosen an inexperienced young man, an intriguer; plainly, I had betrayed the country. The others did not concern themselves with the war; on me must rest the whole responsibility. But Bonaparte triumphs, and then it is Barras who has got him appointed, and to whom alone he is under any obligation."

When Hoche had pacified La Vendée, Carnot had an opportunity of reverting to his favourite scheme of an attack on England, and matured with Wolfe Tone and Hoche the expedition to Ireland, which so nearly proved a success. The victories of the French Republic in 1796 and 1797 far surpassed those of earlier years. By the preliminaries of Leoben, concluded by Bonaparte in April, 1797, Austria ceded Belgium to France, and recognized the Lombard Republic; but Carnot, who was eager for peace, had much difficulty in persuading his colleagues in the Directory to be satisfied even with these conditions. He had been, indeed, on bad terms throughout with the majority—Barras, Rewbell, and La Revellière; and this mutual distrust became aggravated as the Royalist element in the Councils became more threatening to the Directory. Carnot insisted on respect for the Constitution; Barras and his associates decided that the counter-revolution must be anticipated by a *coup-d'état*. When their stroke fell, on the 5th September, 1797, Carnot himself was included among its victims. The soldiers sent to seize him were already in the house when he escaped from it; and after remaining for some weeks in hiding in Paris, he made his way to Geneva in disguise. Even there he was not safe. In spite of his disguise he was recognized by a police spy, and the French Minister at Geneva demanded his arrest. The Government dared not refuse, but sent him private warning. He revealed himself to his hostess, a washerwoman, who made him put on a blouse and a night-cap; and in this new disguise, with a basket of linen on his shoulder, he passed the man who had discovered him, unrecognized, just as the troops marched up to arrest him. After staying at various places in Switzerland he went to Augsburg, where he wrote, in May, 1798, his reply to Bailleul's Report to the Council of Five Hundred, in which he was charged with Royalist sympathies, and persistent obstructiveness. He claimed a fair trial, sure that neither judge nor jury could be more Republican than himself. "My only crime," he declared, "is that I tried to prevent the French people from having tyrants over them. I could not help failing in this attempt, because I refused to make use of any other means than those authorized by the Constitution entrusted to my keeping, to oppose monsters to whom nothing is sacred." Finding himself recognized at Augsburg, Carnot soon afterwards removed to Nuremberg, and remained there until the downfall of the Directory, and the cancelling of its proscriptions, allowed him to return to France. In August, 1796, he had been elected a Member of the Institute, and in acknowledgment he had published his "Reflections on the Metaphysic of the

Infinitesimal Calculus," which had been written some years before. But when he was proscribed, the Institute had been called upon to elect some one in his stead, and his place had been taken by Bonaparte. He was re-elected, however, in March, 1800.

Carnot was soon once more employed in military administration, but under very altered circumstances. On the 2nd of April, 1800, he reluctantly accepted the portfolio of Minister of War under Bonaparte, who had now become First Consul. He feared that he should not be allowed that liberty of action to which he had hitherto been accustomed, and which the state of affairs demanded. In spite of assurances to the contrary, his anticipations were soon realized. He found himself ill-supported, or overruled, by the consuls; and he himself felt a growing distrust of Bonaparte. "This man," he said one day, after an interview with him, "is not straightforward; he wants ministers merely for form's sake, men who will be *his* ministers, not *French* ministers. I shall not be here long." At another time, "Ah! how much good he might have done! His reckless crescendo makes me tremble for him, and much more for France." At the end of September his resignation, already once before tendered, was again sent in, and was accepted. His successor, Berthier, wished to make him a general of division, in recognition of his services, and to place him at the head of the Corps of Engineers; but Bonaparte dismissed the proposal with the note—" Carnot should be nothing under a Republic."

After more than a year spent in retirement, Carnot was chosen a Member of the Tribunate, and there he successively opposed the institution of the Legion of Honour, the Life-Consulship, and the establishment of the Empire.

In favour of the latter, it was said: "That the Republican system had been tried in vain in every possible form; that the only result of all these attempts was anarchy, a prolonged or constantly recommencing revolution, a perpetual dread of new disorders, and hence a general and profound desire to see the old hereditary government re-established, merely changing the dynasty." To this he replied: "The government of a single man is anything but a guarantee of stability and tranquillity. The Roman empire did not last longer than the Roman republic. Its internal troubles were greater, and crime more abundant. The republican pride, heroism, and masculine virtues were replaced by the most absurd superciliousness, the vilest adulation, the most unbridled cupidity, the most absolute indifference to the national well-being. . . . It is not owing to the nature of their government that great republics are wanting in stability; it is because, being improvised in the midst of tempests, they are always built up under the influence of excitement. . . . When one is able to establish a new order of things undisturbed by factions . . . . . it is easier to found a republic without anarchy than a monarchy without despotism."

He stood alone, however, in his speech and in his vote; the Empire was established; the Tribunate was suppressed; and for the next few years Carnot devoted himself chiefly to mathematical studies, and to work in connection with the Institute.

In 1809, his small private means were much reduced by a bad speculation, and he found it necessary to apply to the Minister of War for employment. Napoleon, on learning of this, at once accorded him a pension of ten thousand francs as a retired minister, adding: "He may be useful in many ways; I shall have no difficulty in employing him according to his wish."³

Napoleon was at this time near Vienna, for the campaign of Wagram was in progress; and from that distance he was watching with much anxiety the progress of the British expedition against Antwerp, his new naval arsenal. In the beginning of August, the English troops landed in Walcheren, and prepared to besiege Flushing. On the 22nd, Napoleon wrote to his Minister of Marine:—
"Flushing is impregnable; shells have no effect on a fortress;" but already, on the 14th, the place had surrendered after a forty-two hours' bombardment. On the 2nd September, the Minister of War received instructions from him to have articles inserted in the journals on the cowardice of the general in command, and on the terrible punishment reserved for commandants who should incur such disgrace.

It seems to have been this occurrence that led Napoleon to call on Carnot for a work upon the defence of fortresses, for use at the school of Metz. In a letter to his Minister of War, on the 1st October, he described fully the sort of work that he wanted. It should contain all the orders and regulations bearing upon the duties of commandants, and the sentences passed upon those who have failed to do their duty. It should further show:—

"(1) How true soldiers, taking command of places almost dismantled, have in a short time put them into a state to sustain a long siege. It should enter into full detail about this, and should give a number of instances, such as that of the Duke of Guise at Metz, and that of the Chevalier Bayard at Mézières.

³ Correspondence, June 17, 1809.

"(2) How these brave commandants, anticipating the enemy's attack, have at once made good a breach, or retrenched a bastion, and how, too, the most insignificant work with a good defence has retarded the besieger's progress in the later stages of an attack. The last siege of Dantzig may be quoted, where a simple blockhouse made us spend a fortnight in the crowning of the covered way, and the passage of the ditch. Apropos of this, it is necessary to make a protest against that craze of engineer officers, that a place can only hold out so many days; to show how absurd it is, and to quote well-known instances in which, instead of the calculated number of days for the saps, a much longer time has been required, either owing to sorties, or to some other hindrance due to the defenders; to bring out the resources that still remain after a breach has been made, if the counterscarp has not been blown in, and all fire is not silenced; and to show that even the assault of the breach may fail, if it has been retrenched. This is only an outline," he adds, "of the ideas that should enter into this work; it is an original treatise that is wanted, and I think that Carnot, or some one else of that kind, would be very fit to undertake it. The object should be to make men feel the importance of the defence of fortresses, and to excite the enthusiasm of young soldiers by a great number of examples. . . . It should be a work at once scientific and historical, and the narratives should sometimes be even amusing."

Upon these lines the "Traité de la Défense des Places Fortes" was written. It was originally, as Carnot himself said, a mere "ouvrage de circonstance," which he would not have undertaken had he not himself always held the principles which the work was to advocate.

After filling up the Emperor's outline, however—dwelling on what commandants should do and may do, exposing the fallacies of Cormontaingne's calculations of sieges, quoting official regulations, and giving nearly fifty examples from ancient and modern history—he passed to the principles on which he proposed to rest a new method of defence. It will be best to describe this method, so far as space allows, in his own words.

Vauban's method of attack had "as its object and as its result, not to leave any point upon the ramparts where the defenders could live, or where they could mount a single gun." He afterwards tried at Neuf-Brisach to restore to the besieged, by means of casemates, the cover of which he had deprived them by introducing ricochet-fire; but they only partially attained his object, and merely served to show what had been his intention."

Casemates, abandoned by the immediate successors of Vauban, were once more revived by Montalembert. "In his system of fortification they were no longer a sort of accessory, but the fundamental principle of all his constructions; he proved by experiments on a large scale that the defects which had led to their rejection could be cured, and that it was easy to use them. But the applications which he made of them to the composition of a large number of systems were not happy. He trusted too much to the quantity of his casemated fire to prevent the enemy's approach; he did not screen them from the batteries outside, which he assumed he could silence by his own, heedless of the fact that though he might have the superiority in number of pieces, the keys of the arches, exposed to direct fire, could not fail to be soon destroyed, without possibility of repair; and that the amount of ammunition that would be required to arrest the enemy's progress, was in itself an insuperable obstacle to the adoption of his systems." Still casemates are the only means of protecting the guns and men, for, as he says elsewhere:—"I regard as inadmissible all the machines for successively raising and lowering the guns, so as to be able alternately to load them under cover, and to fire them over the parapet. These complicated machines could not stand against shells, and would be intolerably costly, even if they were provided only for a dozen pieces." (p. 346.)

The question then is, how are casemates to be made use of against the besieger and yet be hidden from him? Only "by substituting curved or vertical fire, such as that of mortars or pierriers, for the direct fire of guns and musketry." This substitution is one of the two main points on which Carnot's method of defence is based.

But it is not in itself sufficient to destroy the efficacy of Vauban's method of attack. "That method consists in moving forward with small numbers, and step by step, closing round and gradually enveloping all the works of the place by lines always well knit together and supporting one another; never forcing matters, if it can be helped; never massing troops at one point; never risking a considerable portion of the army in hazardous assaults, as used to be done before his time."

But in war, as Frederic the Great has said, we should always seek what the enemy shuns. "His main principle is to drive you methodically, and step by step, out of all your positions; yours should be to compel him to make assaults in all cases." If he does this in force, "it is necessary to give way to him for the moment, and leave him exposed as long as possible to a most lively fire from all the neighbouring works, which ought to be in readiness for this purpose. If he obstinately holds the ground he has won, he will lose his whole detachment; if he retires, then is the time to come back upon him in force with the utmost vigour, to pursue him closely without compromising oneself, level his lodgments, and withdraw promptly so as not to remain exposed to his fire, when he has effected his retreat."

But supposing he tries instead to advance according to Vauban's method, then, " if he contents himself with putting a few workmen in his sapheads, sudden sorties of small bodies should be made, to kill the workmen, and destroy their work; and if on the contrary the enemy keeps a large force near at hand to support these workmen, the intended object will have been attained, of drawing the besieger in mass within reach of the heavy vertical fire from the casemates before mentioned."

"The new method of defence consists then in this alternation of sorties and vertical fire; so that the enemy cannot escape the latter without exposing himself to the former, nor guard against them without being overwhelmed by the other."

Not that direct fire is to be entirely excluded from the defence of fortresses. "It is required to oppose the establishment of the first batteries, and to take the enemy by surprise by bringing up guns suddenly, now at one point and now at another: it should also be immediately brought to bear whenever the enemy happens to mask his own fire by his new works; and lastly, it is required to sweep the ditches when the enemy tries to surprise the town or attempts escalade. But these are only momentary opportunities. For the regular procedure another kind of fire is needed, which will search out the enemy in his trenches ; in other words, vertical fire must habitually play the principal part, and direct fire is only secondary."

It is not until the besieger reaches the foot of the glacis that the new method comes into operation. The distant defence and the close defence have widely different objects. "The aim of the former is merely to delay the enemy's progress, that of the latter is to stop him altogether, or to destroy him utterly if he persists in trying to get forward. When the besieger is still at a distance, well intrenched in his camp or his parallels, it would be too hazardous to go and offer him combat with very inferior forces ; we must confine ourselves to harassing him. It would be too wasteful, also, at this distance, to try to destroy his works by artillery fire : all the ammunition which a place could hold would soon be used up without any great effect, for not more than two or three shots out of every hundred would hit their mark." To oblige the enemy to make his works solidly, and to observe all the usual precautions, is as much as it is of any use to attempt at that time, and ammunition should be studiously economized for the final period of the defence.

Carnot anticipates a criticism that has often been made on his proposals, that it is nothing new to employ sorties and vertical fire. "True," he says, "sorties are made and pierriers are provided, but these means are completely secondary in the actual system of defence, whereas they ought to be primary. Sorties are made on a large scale and at rare intervals ; they ought, on the contrary, to be small and frequent, and to ensure their effect there should be a multitude of points of issue upon all the avenues to the place; but there are not. A few pierriers are provided, but there ought to be a very large number; they should be secure from destruction, and for this blindages or casemates are necessary, and there is nothing of the kind. It is half-measures that ruin everything, that throw discredit on what is best, bring all schemes to failure, and only serve to aggravate mischief in all cases."

Carnot proposed his method for the defence of existing fortresses, but the scanty provision in them, alike for sorties and for sheltered vertical fire, led him, in his third edition, which was published in 1812, to consider the alterations which it would be desirable to make on this account in the construction of new places.[4] These alterations chiefly consisted in the construction of a general retrenchment inside the body of the place, and of counterguards outside; in the provision of casemates for flanking the curtains, and for vertical fire; in the detachment of the escarp walls from the ramparts, and in the abolition of counterscarp walls, and of the ordinary glacis.

One ought to be able to make sorties, he says, "when one pleases, where one pleases, and in whatever force one pleases;" but at present nothing is more difficult. "The covered way is protected by a double palisade, which hinders issue even more than entrance, and beyond which there is a continuous parapet, seemingly made only to arrest those who are inside. The besieged is thus obliged

---

[4] Already, in 1797, he had privately circulated a sketch of an angular system, in which the germs of some of his later ideas may be traced, although at that time he had not made up his mind to depend upon vertical, instead of direct fire.—*See* Gay De Vernon's "Traité d'Art Militaire," 1805.

to file through a few barrier-gates, provided in the branches of the covered way, or in the re-entering places of arms. The enemy watches and blocks these narrow outlets; one has to issue from them in his presence, form up before him, and march upon the front of his lines. This lengthy preparation gives him time to get ready for you. First he brings upon you a heavy and deliberate fire, then, as the stronger, he starts in pursuit of you; you have to get back through your defile when you have barely shown yourselves, and if you have not given yourselves time enough you will be cut off. So much for sorties from the covered way." And when the covered way is once taken, how is it possible to recapture it, if there is a counterscarp wall? "Are you to advance up narrow stairs, where you have to mount man by man, to the attack of a superior enemy who is waiting for you, and who, if he were afraid of any return by these stairs, would soon have destroyed them?"

Besides, ricochet fire has deprived the covered way of its old value, and it no longer admits of a stubborn defence, while the glacis absorbs earth that would be most useful for the ramparts, and by the direction of its slope helps to cover the besieger in his saps.

Accordingly, Carnot proposes to get rid altogether of counterscarp wall, covered way, and glacis, and to substitute for them a counter-sloping glacis of very gentle inclination, running up from the bottom of the ditch to the level of the natural ground. This will furnish earth instead of absorbing it; the enemy in sapping across it will be exposed to a plunging fire; and, above all, troops can advance up it to make sorties in any direction, and with any required breadth of front.

The counterguards take the place of the ordinary glacis in protecting the escarp walls from being breached by fire from a distance, and they are so narrow that there is no room for the besieger to make his breaching batteries on them. They themselves have no walls, as they are not intended to be held against assault.

The detached escarp walls adopted for the main works not only have the advantage that, if they are breached they do not bring down the parapet behind them to smoothen the breach, as is the case with retaining walls; but being built with arcades and loopholes, in one or two tiers, they form casemates furnishing a valuable fire. For use in them Carnot invented small hand-mortars, which weighed only twenty-five pounds, and could be worked by a single man, and which could throw 3-inch grenades to a distance of 300 yards at the rate of one round a minute.

Carnot's work at once became famous, and was translated into most foreign languages. But it soon met with severe criticism, which fastened especially upon three points: his exaltation of vertical fire, his countersloping glacis, and his detached walls.

Always too prone to paradox, he had illustrated the value of vertical fire by a calculation of the effect of six 12-inch mortars, placed by pairs at the three salients of a front. Each mortar could at each discharge throw 600 balls of a quarter of a pound weight, and he reckoned that at least one out of 180 balls would take effect, and place a man *hors de combat*. In ten days, therefore, at the rate of 100 rounds a day from each mortar, 20,000 men would be killed or wounded; in other words, in the interval between the opening of the third parallel, and the first assault of the breaches, the whole besieging army would be destroyed.

The fallacies of this calculation have been often pointed out. It makes no allowance for the chances that the mortars will themselves be silenced by the vertical fire of the besiegers, that part of the guard of the trenches will be withdrawn out of range, and that the remainder and the working parties will cover themselves by planks and hurdles against the light missiles proposed. Colonel Augoyat[5] reduced on these accounts the number of men likely to be placed *hors de combat* from 20,000 to 2000. But Sir Howard Douglas and General Eickemeyer seem to have gone too far in asserting that leather clothing would prove a sufficient protection for the workmen against balls which, it is found, will penetrate nearly a quarter of an inch into deal planks, and two or three inches into earth.

Carnot, besides, had no intention of restricting himself to these particular projectiles. He proposed also to make use of shells, grenades, and stones, and indeed of arrows; and those who dispute both the originality and the soundness of his proposals, admit the value of vertical fire, and admit also that it had been unduly neglected. The conditions of the problem have been largely altered by the introduction of rifle ordnance, but in trusting to indirect rather than to direct fire for the defence, he was upon a track on which engineers and artillerymen seem now agreed to follow him. Indeed, the newest idea in fortification, the screened interior battery, finds an anticipation in his pages. He agrees with Marshal

---

[5] "Mémoire sur les Effets des Feux Verticaux."

Saxe in questioning the principle that inner works should always have a command over outer works.

To his countersloping glacis it was reasonably objected that whatever its earlier advantage to the besieged, it would be to the advantage of the besieger as soon as he had established himself on its borders. He would be able to command the gaps between the outworks by which sortie-parties must issue, he would be saved the slow labour of mining his way into the ditch, and it would be far easier for him to assault the breaches or try escalade, than if he had a counterscarp wall in his front. The balance of advantage is, of course, especially unfavourable to small places with weak garrisons; but it must be remembered that Carnot was opposed to such places. "I would abandon them," he says, "that is to say, I would reduce them to mere posts and closed towns, to resist a *coup de main.*" His system is designed for strong garrisons capable of an active defence, and it is inevitable that anything done to allow the besieged to get more readily at the besieger, should also allow the besieger to get more readily at the besieged.

But if the outward communications are to be unhampered by obstacles, they should at all events be well under fire; and here Carnot's dispositions left much to be desired. The counterguards, or as he sometimes calls them, *glacis coupés*, which served instead of the ordinary glacis to hide the escarp walls from the besiegers, served also to hide the besiegers from the body of the place as soon as they had reached the ditch. In this position they were exposed only to the flank fire of some unwalled outworks, from which the defenders could easily be driven away by threat of assault.

But Carnot's faith in vertical fire led him to be increasingly careless about flank defence. When he was preparing a fourth edition of this work, in 1823, he added to it a "Mémoire sur la Fortification Primitive," in which he recommended a simple circular trace, with the body of the place protected by a two-fold envelope, but the ditches of all three wholly unflanked. "It is laid down as an axiom," he says, "that every point should be flanked and defended. But the word 'flanked' is here intrusive; it is quite enough that each point should be defended, no matter how, so long as it is well defended. This may be done by flanking fire; but it may also be done, and often better, by direct fire, curved fire, fire from casemates, sorties, or water manœuvres."

The detached escarp wall is of all Carnot's proposals the one most generally adopted, and most identified with his name, although not really of his invention. The experiment tried against it at Woolwich in 1824, at the instance of Sir Howard Douglas, Carnot's most uncompromising opponent, is interesting as the earliest example of deliberate indirect breaching, but it did not have the discrediting effect which was expected of it. In breaching ten yards of this wall by indirect fire at a range of from 400 to 500 yards, 3436 rounds were fired, of which only 491 struck the wall; and the total weight of iron expended was more than 113 tons, about nine times as much as was required to form a breach twenty-seven yards wide in the escarp of the citadel of Antwerp in 1832, from batteries on the crest of the glacis. It is not surprising, therefore, that in spite of Sir H. Douglas' protest detached walls continued to be built.

Before long Carnot seemed likely to have an opportunity of putting his theories of defence to the test. In 1814, when the allied armies crossed the Rhine and invaded France, he wrote to Napoleon and offered his services. "It is little, no doubt," he said, "that the arm of a man of sixty can do; but it has struck me that the example of a soldier whose patriotic sentiments are well known may rally to your eagles many people who are doubtful how to act, and who might allow themselves to be persuaded that in deserting those eagles they would best serve their country."

Napoleon appointed him Governor of Antwerp, with the rank of General of Division; and on the 2nd of February, he assumed his command. The place was already partially invested by the British and Prussian troops, who on that day drove the French with considerable loss out of the villages of Merxem and Deurne, and on the following day began to bombard the vessels in the docks. This bombardment was kept up for three days, by twenty-four pieces; but the ammunition was insufficient, and effectual measures had been taken for promptly extinguishing fires, so that no great damage was done. The Prussian corps was called away to join the main armies in France, and the British troops under Graham, being only 9000 strong, retired northward. In the following month they made the daring attempt to surprise Berg-op-Zoom which so nearly proved a success, but Antwerp itself was not again molested.

Its governor, however, was active in preparing for a more serious attack. He gathered in provisions, struck a siege coinage of copper to make good the want of money, organized the auxiliary services, and cleared away the faubourgs that would hinder the defence. But his energy was not undiscriminating. The important suburb of Borgerhout lay within the regulation limits for clearance, and the council of defence was of opinion that it ought to be demolished. But Carnot took on himself the responsibility of leaving it, on condition that the inhabitants should put it in a state of defence, and should furnish a battalion to guard it. He had already written in his treatise:—"I look upon suburbs as advanced posts which one may defend for a very long time, and the capture of which, when it occurs, will be of no great benefit to the enemy."

A marble tablet at the entrance to Borgerhout still bears the inscription:—"Au Général Carnot la ville d'Anvers reconnaissante."

Although the place was not besieged, Carnot's position as governor was an anxious one. Of his garrison of 12,000 men, the best troops were called away by Napoleon in the latter part of March, and it became necessary to be constantly on the watch against surprise or treason. Early in April rumours of Napoleon's downfall reached Antwerp, desertions became frequent, and factions began to agitate among the troops. Bernadotte and other leaders of the allies made overtures to Carnot; and Dupont, the Minister of War of the Provisional Government established at Paris, asked for his immediate adhesion to it. But until he received the formal act of abdication of the Emperor he refused to recognize any other authority. On the 3rd of May, he marched out of Antwerp, at the head of his troops, receiving the thanks of the inhabitants, and leaving it to others to hand over the place and its stores to the allies.

He returned to Paris, and at first occupied himself

with scientific studies. But moved by the progress of the reaction, and the abuse poured upon the leaders of the Revolution, he took up his pen to defend them, and wrote a pamphlet which he was forbidden to publish, but was allowed to submit as a "Memorial to the king," and which was soon widely circulated.

In this he charged the returned *émigrés* with being the true authors of the Revolution, the true regicides; and complained that Louis XVIII.—King of France, not King of the French—preferred to owe his crown to foreign bayonets rather than to his own subjects, and had already forgotten so many of his promises. Hereditary right, he reminds him, counts for little with the mass of the people. "It holds that a man has a right to govern when he governs well, and loses this right when he governs ill. . . . We have had Napoleon by the grace of God, and it is by the grace of God that we have him no longer."

However, when Napoleon returned from Elba in the spring of the following year, and offered to Carnot the Ministry of the Interior, he at once accepted it. He had been disgusted with the proceedings of the Royalists; he believed that the Emperor was sincere in avowing his wish for peace, and his intention to govern liberally, and he thought that he himself could promote those objects. When, contrary to his expectation, the allies once more took up arms to restore Louis XVIII., Carnot felt bound to stand by the man who represented the national independence; and he found in the emergency of the situation an excuse for the arbitrary acts which Napoleon soon resorted to. To those who complained of these acts he answered: "Gentlemen, our house is burning, let us all work together to put out the fire, and then you may count on me to help you to repair the building." Napoleon, for his part, was glad of an adherent who had so much weight with the Republicans, and was so strong a guarantee of the good faith of his professions.

Carnot was made a Count, little to his liking, but he was unwilling to weaken the Emperor's cause by openly declining the title. He preferred to leave the official notification unanswered, and his patent unapplied for.

During the few months of his administration he laboured hard in his department, especially to promote primary education. On his first appointment he expressed his surprise that the Emperor should offer him the Ministry of the Interior, instead of the Ministry of War. Napoleon answered: "I naturally thought of that, but your appearance at the War Office would seem an announcement to Europe of my intention to enter upon a great struggle, whereas you know that I am most desirous of peace."

When war had become plainly inevitable, and Napoleon was on the point of joining his army, he explained his plan of operations to Carnot. Carnot strongly urged him not to stake everything upon one battle. The allies would not venture to invade France for another month. Meanwhile he might complete his armaments, and the fortification of Paris; and then, by acting on the defensive, he might protract the campaign till the winter, and gain time to negotiate and divide the allies. "My political position requires a brilliant success, and I shall have one," was Napoleon's reply.

After Napoleon's abdication, the Chambers elected a

Governmental Commission of five members, of whom Carnot was one. On this Commission he did his utmost to enable the capital to show such a front as would impose respect on the allies, and avoid an unconditional surrender; but Fouché, the President of the Commission, had long been intriguing with the Bourbons, and the capitulation of Paris was followed immediately by their restoration.

Carnot's name was upon the first proscription list, and in October, without awaiting trial, he left France under an assumed name and provided with a Russian passport. He never returned, and at the end of the year he was condemned to perpetual exile as a member of the Convention. He went at first to Warsaw, where he was warmly received by the Poles, but he thought it better after a short time to transfer himself to Prussian territory, and obtained permission to reside at Magdeburg. There he lived quietly for the next seven years. He amused himself with writing verses, some of which were set to music by Prieur, his old colleague on the Committee of Public Safety, and became popular in France with many who little suspected their authorship. He also, as already mentioned, made some additions to his treatise on the defence of fortresses, and he had the qualified satisfaction of seeing most of his proposals, though rejected by the corps to which he had himself belonged, adopted by the German engineers in their new works on the Rhine. He died on the 5th of August, 1823.

CHAPTER V.

MODERN FORTS.

The adoption of extended order, which is the most marked change in tactical formations during the present century, is the most marked change in fortification also. Just as the thin veil of skirmishers has grown by degrees into the fighting-line of infantry, so small advanced works have gradually developed into detached forts, and become the true fighting-line of a fortress. And just as military discussion is no longer concerned with the application of the oblique order, but with the handling of companies and battalions in the fight, so it has drifted away from rival systems of fortifying an enceinte to the position and organization of forts. The cause in both cases is the same: increased effect of fire, and diminished apprehension of shock.

It was in field fortification, where there was no great inequality between the combatants, that continuous lines first went out of fashion. Successive improvements in small arms, such as the introduction of the flint-lock fusil, paper cartridges, the iron ramrod, and the socket-bayonet, made infantry stand less in need of a material obstacle to protect them against horse, and allowed of handier formations and greater mobility. Troops on the defensive could not afford to leave the whole benefit of this to their enemy, and restrict themselves to a passive defence. At Pultowa and Fontenoy the example was set of fortifying a position by a chain of redoubts, giving freedom of counter-attack; and the method rose in favour as time went on till its reputation was established by Torres Vedras and Dresden. In permanent fortification the movement in the same direction was necessarily more cautious. Vauban was blamed for the redoubts which he placed on the high ground east of Namur, though they delayed the besiegers for a fortnight in 1695. Marshal Saxe, instead of such redoubts, wished to surround a fortress by a chain of masonry towers. Montalembert combined the towers with the redoubts, and proposed to secure Cherbourg from bombardment by a double chain, of which the outer works were 2000—3000 yards from the place, and 1400—2000 yards apart. But his critics confidently asked what was to prevent a besieger from sapping round and isolating them; for even "when redoubts and lunettes, provided with countermines and susceptible of a good defence, are at the foot of the glacis of a besieged fortress, one does not consider their communication with the place secure unless it is by an underground gallery."[1] Even such a gallery did not save the Queen's Redoubt, in front of Fort St. Philip, in Minorca, from being stormed in 1756, though it allowed the garrison to escape, to surrender with the fort next day. The surprise of Schweidnitz in 1761 increased the mistrust of outlying works. On the other hand, at Cassel, in the following year, "a simple earthen redoubt, 1000 yards from the fortress, obtained all the honours of a regular siege.

[1] "Mémoires sur la Fortification Perpendiculaire."

"The besieger, after spending eleven days of battering and sapping upon this redoubt, at length ventured to assault it; the French, who were defending the redoubt, stood the assault for more than an hour, and at length, with the help of reinforcements from the place, repulsed it with great loss to the enemy."[2]

D'Arçon took part as a young officer in the defence of Cassel, and laid its lesson to heart. In his well-known work, from which the above is quoted, and which was published thirty years later, he proposed the question, "not merely whether detached works should be allowed as an addition, that seems evident enough; but whether, supposing a simple enceinte is to be strengthened, one should not prefer these exterior dispositions to a multiplication of works accumulated upon the enceinte." He laid stress on their importance, as Montalembert had done, for protecting naval arsenals from bombardment, for which purpose they must be thrown forward a mile and a half or more. But he recommended them also for ordinary fortresses which would have little to fear from bombardment, and where they need be only about a quarter of a mile from the covered way. He built, not so much on the delay which their capture would impose, as upon their influence on the *morale* of the garrison, and the activity of the defence. Troops, as he said, were always apt to lose heart and think of surrender as they watched their enemy occupying commanding points, tightening his grip round the fortress, establishing himself on the crest of the covered way, and

---

[2] D'Arçon, "Considerations Militaires et Politiques sur la Fortification."

opening a breach in the body of the place. Detached works, while hindering each of these steps, would also allow the men's courage to be kept up by frequent sorties without risk of their retreat being cut off; and if these sorties were combined with a vigorous use of countermines the defence might be prolonged indefinitely. His Cassel experience had taught him the importance of keeps in the interior of detached works, and of good flank defence for their ditches; and both of these are provided —the former by a loopholed circular tower, the latter by a counterscarp gallery—in the "lunette d'Arçon," of which examples were to be found lately both at Metz and Strassburg.

The detaining value of detached works was exemplified in the two British sieges of Badajoz. The first consisted mainly of an unsuccessful attack on Fort San Cristoval; in the second more than a week was spent in the capture of the Picurina lunette. But the siege of Colberg by the French in 1807 furnished a more striking instance. Gneisenau, then only a Major, was sent there to direct the defence. In presence of the besiegers he threw up a field-work on the Wolfsberg, more than 1000 yards in front of the covered way, to serve as the main pivot of an advanced line. Before it was finished it was stormed by the enemy, but it was recaptured before morning. The French then opened a systematic attack upon it, and after a siege of twenty-five days the work was surrendered, as it had been so much injured by that time that it was unfit to stand another assault. Three days afterwards a sortie was made and the garrison once more got possession of it; but it had to be abandoned next morning, after the gorge had been partially thrown open. Another sortie four days later was repulsed with serious loss, and from that time the defence had to be more passive; but in another fortnight, while the fortress still held out, peace was concluded.

The feebleness of small, old-fashioned fortresses as ordinarily defended, repeatedly illustrated throughout the Napoleonic wars, had been brought out with special prominence in 1814-15; and after Waterloo, a strong current of opinion soon showed itself in favour of fewer and larger fortresses. Rogniat led the way in 1816, with a proposal to convert places of strategic importance into intrenched camps capable of receiving an army of 100,000 men by adding four forts, one on each side of the place, and a mile and a half from it. Four battle-fields would thus be presented to the enemy, each about three miles long, perfectly secured on the flanks, and strengthened by field-works in the centre. The problem of fortifying Paris gave practical interest to the question, and after twenty years of controversy between the advocates of detached forts and those of a continuous enceinte, it was decided in 1840 to provide both.

Meanwhile, in Germany, this solution was more promptly and unanimously arrived at. The most important fortresses on the Rhine, Cologne, Coblentz, Mayence, and Rastadt, were converted into intrenched camps by the help of detached works, and the same thing was done with Verona, Ulm, Olmutz, Cracow, and the places on the east frontier of Prussia; while at Lintz an intrenched camp was formed by a chain of towers without any interior enceinte.

The character of the German works and the principles

on which they were planned were explained in a memorandum by General von Brèse, the designer of the works of Posen.³ When a position is to be fortified, the engineer, he says, makes choice of the most important points, and on these he places round or angular towers, forming casemated defensible barracks of two or three stories, with a gun-platform on the top. These towers, secure in themselves against a *coup de main*, are sheltered from direct artillery fire by earthen ramparts in front of them, upon those sides on which an enemy could place batteries. Thus they form the keeps of works, sweeping the interior with musketry from their lower story, and the terreplein of the ramparts with artillery from their upper story, while the gun-platform on the top commands the country as a cavalier. The upper story can also be used for howitzers to shell the siege-works. The ditch in front of the earthen rampart is flanked by caponiers or counterscarp casemates; and, unless the work is exposed to fire on all sides, the gorge is closed merely by a wall, so that other works in rear may fire into it if it should be taken by the enemy. The Cologne works of this type were placed about a quarter of a mile in front of the enceinte, and rather more than that distance apart.

A chain of such works, General von Brèse argued, is not only cheaper in construction, garrison, and armament, than a corresponding length of bastioned enceinte, it opposes a resistance three or four times as great, since each work must be taken separately, and above all it gives

---

³ Extracts from this memorandum were given in the "Royal Engineers' Aide-Mémoire to the Military Sciences."

every facility and support to the offensive strokes of the garrison.

As soon as rifled artillery was introduced, it was seen that the great increase of range and accuracy made detached works at once more necessary and more effective than they had been hitherto, and they were employed on quite a new scale. In 1859, immediately after the peace of Villafranca, Verona was provided on the west side with an outer chain of forts two miles in advance of the enceinte, and rather more than one mile apart. The new fortifications of Antwerp comprised a chain of forts which were similarly placed, but were of much greater size, having a crest line of nearly half a mile on the front and flanks, and an armament of 120 guns.

In the English defence works undertaken at the same time, they played a still more prominent part. "When the extent of the positions necessary to be occupied in order to protect the dockyards against long-range bombardment is considered, it is evidently impossible," it was said, "to occupy them by continuous lines, which must be manned throughout their whole extent, and which fall if pierced at any one point."[4] The forts in some cases had to be thrown forward three or four miles, so that they must needs depend wholly upon themselves, without any support from an enceinte in rear. They were rather nearer to one another than the Antwerp forts, and had only about half the length of crest-line. While of the same general type as the German forts, more care was taken to place their caponiers in such positions that they were not liable to be silenced by fire along the ditches they flanked.

[4] "Royal Engineer Professional Papers," vol. ix.

In England, as before in France, the new fashion of fortifying did not meet with unanimous approval, in spite of the fresh arguments in its favour. Just as conservative soldiers, even after the Franco-German war, still refused to admit that "our two-deep line formation, so long regarded as a thoroughly British institution, must be looked upon henceforth as impracticable, and that the German skirmisher-swarm formation must take its place," so others ten years earlier refused to admit "that mere fort-building is fortification," and asked for "a good wholesome wall and ditch which the enemy has to get through or over before he reaches his object." A very able officer—Colonel Owen—tried to show that a continuous line, which "has been well and fully tried for thousands of years," is as cheap as a line of detached works; that it can be defended by fewer men and those men far less trained; that its defence is simpler and easier understood by generals, by officers, and by men; and that it appeals most to the patriotism of the citizens.

But his vigorous argument failed to win much support, and in the discussion which he started the weight of authority was all against him.[5] It was generally agreed that to guard a very extended line with a small force, the essential thing was to make sure of the commanding points. If these were strongly fortified and held, little was to feared from an enemy pushing between them. Connecting lines would be of more or less value according to circumstances; but a simple continuous line

[5] "Royal Engineer Professional Papers," vols. xii. and xiii. Another advocate of continuous lines has lately come forward, but only to follow in Colonel Owen's track. See "Occasional Papers of the Royal Engineers' Institute," vol. iv.

of many miles would be as ill-adapted to passive defence by a mixed garrison as to active defence by an army of disciplined troops. As regards the relative expense, there was the broad fact that the enceinte of Paris, though of the simplest kind, without outworks and with an unrevetted counterscarp, had cost twenty-five per cent. more than the chain of forts, the perimeter of which was more than half as much again.

The real gist of the question is well brought out by one of Colonel Owen's illustrations. " When a farmer puts up a fence or a wall round his garden, he erects a fortification; and so long as fortification is a development of that one simple idea, every one can understand it. When you put up a row of forts and say the enemy cannot pass through, it is asking the farmer to believe that the posts of his fence will keep out cattle without the rails." But the parallel shows how completely the engineer may lose sight of the tactician. The fence has to keep out the cattle by itself; but it is not the works, but the troops inside of them that will have to keep out the enemy. The question is, how they will be best posted for that purpose: in a long thin line, or in compact groups? The armour must be made to fit the man, not the man the armour.

Colonel Owen took exception, not only to the open intervals, but to the closed gorges of the forts. He confidently asked whether, supposing an engineer had the time and means to strengthen further a place already fortified by a continuous line, he would intrench the salient points inwards, or in other words, convert them into forts. "Surely not," he said; "would he not rather make coupures in his own rampart, and retrench against the enemy those points most liable to be breached, or where from exceptional circumstances an escalade might be apprehended?" The course of the American Civil War was just at that very time throwing some light on this point. General Abbott, of the United States Engineers, in his remarks on the operations against Richmond, dwells particularly on the advantage derived from insulating the vital points of the line. "Our system of intrenchments at Petersburg," he says, "consisted in general terms of a system of field-works, each capable of containing a battery of artillery, and a strong infantry garrison. These works were closed at the gorge, were protected with abattis and palisading, were often supplied with bomb-proofs, and were located at intervals of about 600 yards, on such ground as to well sweep the line in front with artillery fire. They were connected by strong continuous infantry parapets, with obstacles in front." Early one morning, three Confederate divisions swept across the lines on either side of one of the weakest forts, joined in rear, and carried it; but they could take no others. As soon as daylight would permit, all the artillery that could be concentrated opened on the work which had been taken, and which the enemy still held. "No reinforcements could join him from his own lines, owing to this fire which swept his communications; his captured position was entailing deadly loss; our reserves were rapidly assembling, and finally, about 8 a.m., they made a charge which resulted in the recovery of our works, of all our artillery, and in the capture of over 1800 prisoners."[6] The result was very different in some of the

---

[6] Abbott, "Siege Artillery in Virginia."

P

assaults upon the Confederate lines, where the works had open gorges, and such an extent of line was consequently occupied, that the assailants could not be driven out again.

What applies to field-works applies equally to permanent or provisional works. German engineers, from Von Brèse to the present day, have laid it down that in most cases an enceinte will be best formed by placing works of the same general type as detached works upon the most important points, and connecting them with one another by simple lines; and this was the mode adopted in the bridge-head of Florisdorf, thrown up in 1866, to protect Vienna.

The varied siege experience of the Franco-German War brought out fresh reasons for extending the circle of defence of fortresses, both large and small, and consequently for relying more upon forts. The blockade of a fortified capital, like Paris, proved by no means so vast an undertaking as had been predicted. Instead of requiring five or six times the strength of the garrison, as Colonel Brialmont, for instance, had estimated, it was found that a garrison of 200,000 regulars and mobiles (besides national guards) could be securely invested by an army actually inferior to it in numbers, extended over a circuit of nearly fifty miles.

The chief object of the new fortifications of Paris is to make such an operation more difficult in future by robbing the enemy of those commanding sites which proved so useful to him, doubling the length of his line of investment, and embracing large areas of camping ground and pasture. There are three principal masses of high ground, upon the north, the east, and the south-west of Paris. Possession of these masses is now secured by the new forts, which form three separate camps, "three tactical centres," and are from six to ten miles in advance of the enceinte. How far more formidable a work another siege of Paris would be, is fully recognized by the Germans themselves. A writer in the "Militair Wochenblatt" (August, 1880) points out all the difficulties of the task: from the extension of the works, their advantages of position, their excellent construction, and powerful armament, the circular railways which give the commander-in-chief a freedom of action without a parallel in military history, and the immense resources of the city. "One cannot expect that, as in 1870, the French armies will disappear from the scene, but must rather assume that there will be time enough for a large army to be collected for the defence of Paris," especially when we consider the spider's-web character of the French railway network. While a continuous investment would be a gigantic task, he thinks it very questionable whether, as has been proposed, an effective blockade could be maintained by armies concentrated on different sides, and linked to one another by cavalry divisions. But "the defence of this vastest of all fortresses must be planned and executed upon a grand scale, and requires military genius of the highest order," and the doubt whether this would be forthcoming is the chief consolation which the writer offers to his countrymen. But, as the same writer adds, "it is not only round Paris that the observant German soldier sees the circle of defence growing wider and stronger; besides the new intrenched camps of Epinal and Belfort, Langres, and Besançon, the lines of La Fère, the fortified position of Rheims, the fortresses of Verdun and Toul, girdled with strong new forts, the fortified plateau of Haye, the permanent works which guard the Moselle near Nancy, and those which lie on the Meuse, upon the north-east frontier, the entry of an enemy in another direction is opposed by the intrenched camp of Dijon in the Côte d'Or, and by that of Lyons further south. If the above-mentioned works of defence and lines are only in part new creations, yet it is solely by the extension they have received, their solid construction and suitable armament, that they have become factors full of importance, which must be taken account of in those large calculations on which hangs the weal or woe of nations. One stands amazed when one considers that that same France which lay so low in 1871 is now able to call out much more than a million of men to defend their country, and that all the above-mentioned measures of defence have been carried out, with a silence quite unlike the French wont, and are now nearly finished."

What chiefly concerns us here in these defences is to notice the almost complete disappearance of those small fortresses for about 5000 men, which, from Vauban's time down to 1870, stood like picquets extended along the French frontier. Instead of them we see a series of camp-fortresses [7] with barrier-forts between them. Mont médy is to be reduced, it is said, to the latter class; while Toul, Langres, and Verdun have been converted into camp-fortresses of ten times their former diameter. Toul proved very useful in 1870, blocking a main line of railway for six weeks, owing to the German inability to

---

[7] The Austrian term "camp-fortress" seems preferable to the German term "fort-fortress." One can speak of the component parts as camp-forts, instead of as forts of a fort-fortress.

bring up a siege-train at that time; but, like Verdun, Thionville, and so many other places, it surrendered without waiting to be regularly besieged. The convergence of fire from all quarters, to which their small radius exposed them, made defence hopeless. The long and successful resistance of Belfort was mainly due to the bold mainte_ nance of outlying positions. Apart from the reasons that have been given already, the favouring of an active defence, the occupation of valuable sites, the protection of buildings from bombardment, and the enhancement of the difficulty of blockade; looking merely to the conditions of the artillery combat, rifled guns have made it all-important to enlarge the circle of defence, so as at any rate to escape reverse fire.

In some cases, as at Antwerp and Strassburg, the nature of the ground is likely to forbid complete envelopment, and there may be fair hope of reaping the full value of a strong enceinte by an obstinate defence at close quarters; but elsewhere, though it may be well to have some continuous obstacle to bar entry into the town which forms the heart of a fortress, it may be doubted whether it will be found worth while in future to create a siege-enceinte. What used to be said of the covered way—"covered way lost, all is lost"—may now be said with equal truth of the line of forts. That is the line which it is the most essential and easiest to defend, and nearly all the available resources should, therefore, go to reinforce it. The idea of the Prussian engineers of the last generation, that a fortress, like a spiral spring, should offer more resistance the more it is compressed, must be discarded as hopeless in presence of rifled artillery.

Some nucleus (*noyau*) is certainly desirable for every

large fortress, but this may be not in its centre, but on its border-line, like the citadels of former days; some particular region of marked natural strength or importance being constituted as an independent camp, defensible on all sides. The small camp first formed at Langres may be said to stand in this relation to the much larger camp since formed by the new forts to the north and east, two of which are seven miles from the place.

General Brialmont goes further. He recommends that the plan of grouped camps which has been actually adopted at Paris should be the normal disposition for the fortification of a capital. He gives as a type three similar camps symmetrically placed on a belt nearly six miles wide, and lying at about the same distance outside the city. Each camp would be roughly about fourteen miles by six, and might be formed by ten forts—five upon the front, three on the gorge (towards the city), and one at each end. The intervals between the camps would be about nine miles. He considers that though the cost would be greater than that of a simple line of forts, the defence would be much more protracted, and there would no longer be any necessity for a guard-enceinte round the capital, a thing most important to avoid with a vast and growing city like London.

"There is a bold and tactical air about this method of fortifying; holding the enemy off by threatening his flanks instead of barring his front. It looks at first sight like a larger application of the principles which have substituted detached works for continuous lines. But there is the all-important distinction that such works are assumed to be within range of each other, and that these camps are not. The combat is to be one of weeks and not of hours, so that the enemy can afford to push his way forward cautiously in the intervals upon a front of six miles, intrenching himself against flank attacks. Once within range of the capital it must fall into his hands, for any show of resistance would bring bombardment upon it. The defending army, if the war is not ended, will find itself cut in three, and its camps will be open to attack on whichever side the enemy may prefer. It may be said that the intervals would be occupied by field-works; but the comparative inadequacy of field-works, which is the ground for all permanent fortification, applies here as much as elsewhere. No doubt, with so extensive a circuit it would be well worth while to retrench some one part to serve as a citadel; but leaving one of these camps as it stands for this purpose, one cannot help thinking that the half-dozen works which guard the rear of the other two camps would be better placed by pairs in the intervals." [8]

However well such a disposition may apply in a particular instance, such as Paris, where there was not only an enceinte for the city, but gorge forts for the camps ready to hand, it seems very questionable as a general type. As Brialmont himself points out in another connection:—"The action of a fort on the ground in front of the neighbouring forts will be greatest where the forts are in a straight line, or in one that is very slightly convex. One should avoid as much as possible, therefore, placing the forts in such a way as to form pronounced salients and re-entering angles."

With the exception of the works at Paris, the forts

---

[8] *Royal Engineers' Journal*, March, 1874.

lately built have, as a rule, been placed within three miles of the enceinte. It has been recommended by Brialmont [9] and Brunner [1] that this distance should be increased to four miles and a half, in order absolutely to preclude bombardment of the town. But the increase in cost of construction, and in the strength of garrison, involved by such an extension of radius, has to be set against this advantage. Also, as the distance widens, the area of intervening ground unseen from either forts or enceinte, and affording shelter to an enemy who has passed between the forts, will become greater; in one of the new French fortresses there is said to be more than a square mile of such ground. Such an advance will sometimes be necessary, however, in order to see the ground in front better, as on the south-east side of Verdun (which is probably the fortress just referred to), where one of the principal forts is more than four miles from the enceinte ; and occasionally it may even reduce the length of line to be defended, as in the Anthony position at Devonport.

It is beginning to be accepted as a principle that the distance of forts from one another, instead of one mile or one mile and a half, may be two miles and a half or three miles, if the ground is open ; in other words, that it is enough for them to defend the intervals between them instead of affording effective mutual support. This has been acted upon in many of the new fortresses, especially in France. Between the St. Cyr and Palaiseau Forts (on the south-west of Paris), which are more than ten miles apart, it has been thought sufficient to provide two others. The country round Verdun is broken and wooded, and

[9] " Défense des États," (1876), pp. 141, 143.
[1] " Beständige Befestigung," (1876), p. 154.

there eleven works have been made for a perimeter of twenty-five miles, the intervals varying from one mile and a half to three miles. One reason given for this wide spacing is to economise garrisons, and to avoid breaking up the defence into many fractions. But the duty of guarding intervals of three miles could not fail to be made lighter by the existence of storm-proof posts in the middle of them; and the writers who give their sanction to such long intervals (e.g., Brialmont, Brunner, Von Bonin) assume some such small intermediate works, either permanent or provisional.

"It is well known," says a French engineer,[2] "that in the fortresses as they now stand, both in France and abroad, the extent of the works has been limited by financial considerations; and, therefore, in most cases only the most important points round the fortress have been fortified, and the completion of the defences has been postponed till the time of need. . . . The Germans have placed the forts thrown up round Strassburg[3] and Metz at average intervals of two to four kilomètres, as they consider those places to be too much in front line to allow the making of intermediate works to be put off till war breaks out: on the other hand, for the forts of their places in second line they have been satisfied with the intervals of five or six kilomètres' adopted in France. This shows that they mean to throw up intermediate works there when needed, and that they admit the

[2] "Étude sur la Fortification semi-permanente." Par un Officier du Génie. (1880.)
[3] Strassburg has fourteen forts; those on the south and east are four kilomètres, but those on the north-west are only two kilomètres apart.

principle of mutual flanking at effective ranges, a principle which is equally admitted in France."

The projects for such intermediate works have, according to this writer, been already got out, but he estimates that 500 or 600 men would be occupied two or three months in making one, and reasonably urges, therefore, that France should follow the example of Germany, and build them in time of peace for her frontier fortresses.

But if it is desirable that they should be permanent works, one may further ask whether the maxim does not hold good here as elsewhere, that "a chain is no stronger than its weakest link," and whether—apart from peculiarities of site, which will continually give more importance to this fort, and less to that—it should not be the general rule to equalize the several links, instead of making them weak and strong alternately. "The garrison of a large fort consisting of 1500 to 1800 men, commanded by a colonel, will usually have a better spirit and be more ably handled than that of a small fort consisting of 300 to 400 men, commanded by a captain or a major."[4] On this ground General Brialmont has always been the advocate of large forts, and prefers to reduce their number rather than their size; but whatever weight this argument may have against small forts, it must have much more weight against the still smaller intermediate works (intended for fifty or sixty infantry and three or four light guns, according to General von Bonin), which it is proposed to place in the wide intervals between the forts. The French engineers lean to Brialmont's views. The principal forts of their new camp-fortresses have infantry garrisons of 1000 men and mount thirty-six guns

[4] " Défense des États," p. 143.

on the ramparts. Some of the Paris forts mount sixty guns, while some of the isolated barrier forts mount from eighty to 100, and are constructed for 2000 men or more. The German engineers lay less stress on size. They are satisfied in most cases with infantry garrisons of 500 men, and with twenty guns on the ramparts. When more guns are needed at any particular point, they can be placed in wing batteries, without enlarging the forts. If the Germans are right, forts may be placed at a mile and a half apart without absorbing more than one-third of the entire garrison of the fortress, reckoning this, as is generally the case, at about 1000 infantry per mile.

Passing now to the construction and organization of individual forts, it will be best first of all to borrow a general description of one of the new German forts,[5] such as those of Strassburg, to serve as a point of departure and comparison for others.

"The detached forts of the chain round a great place of arms are usually of lunette shape, with very obtuse salients (130° to 145°), in order that the faces may escape enfilade, and also that their frontal action may be better, since they have to hold their own against the besieger's batteries in the first encounter. The direction of the flanks depends on the position of the collateral works, but is, as a rule, nearly parallel to the capital. Assuming the forts to be a long way from the enceinte, their gorges may be closed by a line of rampart either of bastioned or of slightly re-entering trace.

"The size of the forts is proportioned to the part they have to play in the general system of defence. The faces will commonly be seventy-five to 125 mètres, the flanks

---

[5] Bonin, "Festungen und Taktik des Festungskrieges" (1878).

fifty to seventy mètres in length, so that the perimeter of the forts will considerably exceed what it used to be, just as their present importance does.

"In a detached fort the use of the covered way is not so much to facilitate sorties, which can be made more conveniently upon the flanks, as to allow of keeping sentries outside the work up to the last stage of the defence; it can be replaced, therefore, by a simple patrol path, which on the gorge side becomes a roadway to the wing batteries (*Anschluss-glacis*), and widens out opposite the gorge gateway into a place of arms, with a tambour and a block-house to guard the communication. By this arrangement of the covered way, together with admissible reduction in the width of the ditch, it becomes possible to screen the masonry escarp of the faces and flanks from the besieger's indirect fire, provided it is limited to a height of about five mètres. Since this height in itself affords no sufficient security against assault, and the situation of the forts—usually on commanding ground—seldom allows of wet ditches, such security must be sought for by means of good flank defence for the ditches, and of a high revetted counterscarp, perhaps organized for counter-mines. The escarp of the gorge, not being exposed to the heavy batteries of the attack, will be utilised for shelter-casemates, and revetted to a suitable height.

"The double caponiers at the shoulders which were formerly used to flank the ditches of lunettes, are no longer admissible, as they could be destroyed by the enemy's indirect fire on the prolongation of the ditches of the faces. They are usually replaced by a caponier at the salient, sweeping the ditches of the faces with artillery, and two single caponiers at the shoulders, on the prolongation of the escarp of the faces, sweeping the shorter flanks with musketry. If the gorge has been given a bastioned trace, casemated flanks furnish a low flank-defence for its ditch; if it has been simply broken inward, a caponier is required. All caponiers have posterns leading to them from the interior of the fort.

"The old reason for providing the escarp with a costly revetment wall, to bring the musketry fire as near as might be to the crest of the glacis, has no longer any weight with the new rifle; it is now thought better, therefore, for the sake of economy, to have an earthen slope for the escarp of the rampart of the faces and flanks, and to place a less substantial detached wall at the foot of it as the obstacle to assault.

"The command to be given to the rampart above the plane of site depends upon the formation of the ground in front, which should be overlooked as extensively as possible; it will seldom be less than eight or nine mètres. The greater penetration of the new siege-guns requires that the thickness of the parapet should be increased to about seven mètres. A less command will suffice for the gorge parapet; it is enough that it should cover the faces and flanks from reverse fire in case of assaults; and here, as the direct fire of heavy siege-guns has not to be met, the thickness of the parapet may be reduced to four mètres.

"The rampart will from the first be organized for guns upon the faces and flanks. On the gorge it will be prepared only for musketry, but in the later stages of the defence it may be necessary to mount guns there also; and accordingly the terreplein must be made wide enough for them. . . .

"The employment of casemated batteries for fire to the

front has had to be given up; the bomb-proof gun emplacements of masonry or timber which were formerly in use, can be used no longer, owing to the accuracy and destructiveness of the new siege artillery; it has been found necessary to sacrifice the howitzer fire upon the country which was to be furnished by the upper stories of casemated keeps, since such action of the keeps could not be reconciled with their own protection from indirect fire. But this protection has further required in most cases that the command of the interior of the work from the casemated keep should also be sacrificed, since if it was to fulfil its object in this respect it could not be shielded against an angle of descent of 10°. There was naturally great reluctance to abandon the casemated defences which had been thought so much of, and attempts were made to reinforce them by iron plating, as had been successfully done with ships of war. But the results of these attempts were not very satisfactory. The conclusion has had to be accepted that casemated defences are only to be employed where, owing to the nature of the site, they can be protected even from indirect fire. · . . .

"The importance of the rampart as the main line of defence has become more distinctly marked, and has demanded better measures for securing its defensive powers. With this object the gun emplacements and roadways along the ramparts have been sunk lower, in order to get better cover; higher carriages have been adopted for the guns firing overbank to repel assaults; while indirect fire is chiefly contemplated from those which are intended for the artillery combat in case of a regular siege, since the deep embrasures that have been used hitherto cannot now be employed, on account of the accuracy of the new artillery. More complete traverse-cover has been given to guard against side shots; only one gun being placed between a pair of traverses on lines exposed to enfilade, and on other lines at most two guns. Secure receptacles for ammunition for immediate use have been made in the interior slope, or in the traverses; and the latter, being built hollow, with a sufficient thickness of earth over the masonry on the enemy's side, allow the troops on duty upon the ramparts, and even some of the lighter guns provided against an assault, to take shelter in them if the enemy's fire should be very heavy. . . .

"The quarters for the bulk of the garrison lie as a rule in a range of casemates along the gorge, which also contains the hospitals (essential for the self-dependence of the several forts), and in the basement, kitchens, stores, &c. For guards and other detachments held in readiness, casemates open to the rear and well provided with outlets are made under the terreplein of the faces; in case of need the space afforded by the posterns will either supplement these or serve instead of them. This space will also be sufficient to contain some of the stores which are not immediately required for the defence, and which ought to be sheltered from the enemy's fire. . . .

"Each fort requires at least one main powder-magazine, completely in the heart of the rampart, and quite out of reach of the enemy's fire, and, according to its size, two or three shell-filling rooms, each with its own expense-magazine and the necessary storerooms for the several kinds of artillery ammunition. . . . . These are placed under the terrepleins, and are connected by lifts with hollow traverses overhead, so that the made-up ammunition can be transported without risk from the enemy's fire almost to the very spot where it is to be used. Lastly, the increased effect of siege artillery imposes greater care about the communications. This is partly met by the larger use of traverses on the terrepleins; but with .... detached works one cannot do without large capital, or central, traverses, which divide the interior of the work into separate portions, protect the several lines of rampart from indirect reverse fire, and at the same time allow a completely sheltered communication to be made between the gorge and the casemates under the main rampart. With very exposed works one may even go so far as to carry this covered communication right along the ramparts, and connect it by staircases with the hollow traverses on the terreplein, so that if the enemy's fire is heavy and convergent the open area of the work and the ramps to the terreplein need not be used at all for circulation."

Much of the above description would apply to any of the recently constructed forts, whether in Germany or elsewhere. The points which chiefly call for remark, and regarding which we find most difference of treatment, are :—

(*a*) The general shape of the fort,

(*b*) The provision of a keep,

(*c*) The disposition of the rampart armament and the mode of mounting it,

(*d*) The caponiers.

General Brialmont prefers a single straight front, or head, to the two faces of the more usual lunette form. His objection to the latter is that they bear less directly upon the ground over which the besieger will advance, and that they are likely to be enfiladed from the salient towards the shoulders. When a work occupies a re-entering position in the general line, so that the prolongations of the front ditch cannot be taken up by the enemy, the straight trace has everything to recommend it. For one thing, its caponier may be placed at one end instead of in the middle, and will be able to defend also the adjoining flank, thereby saving a shoulder-caponier. This will also allow such a trace to be adopted in cases where one only of the prolongations lies out of the enemy's reach. But Brialmont himself recognizes that the front must be broken outwards whenever the enemy could otherwise place batteries so as to breach the caponier by firing along the ditch. The forts of a camp-fortress must as a rule be salient works, and their saliency increases as the intervals between them are made wider. Even if the siege batteries are not strictly on the prolongation of the ditch, but 10° or 15° outside of it, their shot will still drop sufficiently after clearing the glacis to strike the masonry of the caponier. So far as the escarp line is concerned, therefore, it seems likely that as breaching by curved fire becomes more perfect the angle between the faces, instead of being increased to 180°, will commonly have to be made less obtuse than hitherto; and if this makes the fire of the two faces too divergent, or exposes them too seriously to enfilade in the opposite direction, some part of them must be traced independently of the escarp.

In some cases secondary flanks are provided, firing to the right and left rear, so that the work fronts three-quarters of a circle, and the gorge is narrowed. In very salient positions such a form is imperative, and even for the ordinary forts of a chain it has the advantage that, when their neighbours have fallen, and the enemy pushes in through the gap, they themselves cannot so easily be taken in rear, and are better able to support a retrenchment line. But these secondary flanks are themselves so exposed to reverse fire, that they need to be protected by parados, or to be completely casemated. They help to hide the keep, when there is one, from the enemy's view; but that seems a questionable advantage, as they correspondingly restrict its action.

The new French and German forts have been made without keeps, on account of the difficulty already mentioned of making them effective for their purpose, and at the same time sheltering them sufficiently from curved fire. "They restrict the interior space," says Major Brunner, "and intercept the shells which fly over the parapet of the work, so that they are liable to be disabled along with it Consequently their services are not always in proportion to their cost."[6] Another Austrian writer says: "Although keeps must be protected from curved fire in the same way as the main work, it must be remembered that the keep has usually no action upon the ground outside, and that if once the outer line of the fort is carried, the pushing forward of an attack step by step on the keep, though it presents difficulties and occupies men, causes no serious hindrance to the progress of the attack against adjoining works, or against the nucleus."[7]

But other writers still insist on their importance. "To renounce a keep," says Wagner, "would be as in-

[6] "Beständige Befestigung," p. 89.
[7] Weeger and Geldern, "Befestigungskunst" (1873), ii. 36.

correct as to fight in the field without reserves." [8] He endeavours to adapt the old Prussian type of keep to present conditions, retaining even the casemates for high-angle fire. Instead of over-lapping the gorge ditch, it must now be pushed forward into the interior of the work, and its escarp covered by an inner glacis. Its platform must not be given any command over the main rampart, but one or more iron turrets may be placed on it to fire indirectly upon the country.

Brialmont maintains that "a keep is an indispensable work for every important fort which is either isolated, or in a salient position." [9] It is not really costly, for it furnishes the casemated cover which must needs be provided in some form, and by the additional security it gives, it may even allow a reduction in the height of the main escarp and counterscarp. The space it occupies is useless for any other purpose, and it does not necessarily involve any enlargement of the fort. If properly constructed, instead of falling with the fort, it will impose a second siege on the enemy, and will meanwhile arrest his further advance; for, unlike Wagner, Brialmont makes it an essential condition that the keep should overlap the gorge ditch, and while screened from the front by the main rampart, should be able to command the country on either flank as well as to the rear. He dwells upon the value of the keep, not only in guarding a fort against sudden capture by assault, but also in supporting counter-strokes for its recovery when the enemy have got possession of it. With this contingency in view, he would make wide roadways across the gorge ditch and

---

[8] "Grundriss der Fortification" (1872), § 129.
[9] "Défense des Etats," p. 185.

rampart, close under the guns of the keep, to allow the fresh troops to enter the work.

Since he wrote, a marked instance in support of his argument has been furnished by the storming of Kars. When the Russians had made themselves masters of Fort Kanly, the fire from its keep, a defensible barrack, obliged them to abandon the interior, and get cover outside the parapet. It enabled the Turks to reoccupy the work for a time, and when they had retired the keep still held out, until all the works on that side of the river had surrendered, and it was plain that further resistance was useless. It must be remembered, too, that Fort Kanly, like the other southern forts, had been vigorously bombarded with siege-guns for a week before the assault.

The main problem in fort building—how to organise the ramparts and dispose the armament, so as to deal either with the converging fire of numerous siege-batteries, or with sudden assaults by overwhelming forces—meets with widely different solutions. In most of the German forts, as already mentioned, reliance is placed chiefly upon the use of high carriages, and of massive and frequent traverses. There is only a single rampart, on which the heavy guns are mounted for indirect fire during the artillery combat, and which serves for infantry, or for the lighter guns firing overbank when the enemy comes near. In the French forts there are two ramparts, a lower one for infantry, and a higher one behind it for artillery, so that 600 rifles and about thirty guns can be brought into play at once. This arrangement is said to have been adopted also by the Germans in some of their latest works, but Brialmont condemns it because the infantry would suffer so much from shells bursting in the exterior slope of the rampart in their rear. He himself recommends an inner rampart for the heavy guns, but it is lower than the outer rampart, and screened by it from the enemy's view.[1] In large forts with keeps, this inner rampart should be broken into two halves, separated by the keep, which should flank them both in front and rear; in small forts it will form a continuous retrenchment or parados for the gorge parapet.

An interior battery of this kind has been adopted in the latest English type of fort.[2] The crest is about sixty yards behind the outer parapet, and half a yard below it, and the exterior slope is very gentle, so that shells striking it may ricochet over the work. The outer rampart is intended only for musketry and field-guns, with a few heavy guns on Moncrieff carriages at the angles.

In the newest Austrian designs, given in the magnificent collection of details of military architecture recently published,[3]—a collection to which one would much like to see some English parallel—we also meet with interior batteries, with their crest on the same level as the outer crest, or a few inches higher. In one case the battery forms part of a keep. The ends

---

[1] General Todleben carried out some experiments in 1875, which led him to the conclusion that, as indirect laying must in any case be largely employed in future, part of the armament should be placed behind the main rampart instead of upon it.

[2] "Royal Engineers' Institute Occasional Papers," vol. vii. (1882).

[3] "Sammlung von Constructions—Details der Kriegsbaukunst lithographirt im k. k. t. und a. Militär-Comité." Wien, 1880.

have a masonry escarp screened by an inner glacis, and they form orillons covering the flanks of the keep, which have a good view over the country upon the sides, and in rear of, the fort. The front has only a steep earthen slope, with an unflanked palisade at its base.

Iron turrets and shields have been made use of to some extent abroad for inland works as well as for coast batteries. The earliest turret mounted on land was on the keep of one of the Antwerp forts, and some of the new forts at Metz are provided with two turrets each. The cost has been brought down as low as 4000*l.* for a two-gun turret; and Brunner points out that a fort with ten such turrets will even be cheaper than a fort for twenty unarmoured guns, on account of its much smaller size. But the uncertainty of artillery progress makes it unsafe to stake much on them, and so they stand on the footing of defensive luxuries, additions to works which are not dependent upon them, but for which no money is grudged. It is said that there are not more than a dozen turrets in the whole of the fortresses of Germany. They are chiefly to be used, according to Wagner, for giving protection from curved fire to heavy guns which are hidden from view, and are themselves to be laid indirectly, as on the keep of the fort referred to above.

In the French forts iron has been more largely used, shields being provided for several guns on the ramparts. Shields give a much more restricted field of fire than turrets, and their ports cannot be averted from the enemy when not in use; but they cost much less, and it is easier to increase their thickness if it should become necessary at any time, supposing that they are made of plates bolted together. It is a great drawback to the chilled cast-iron (Gruson's patent) which is now being so largely adopted abroad, both for shields and turrets, that it will hardly admit of any such subsequent strengthening.

The flank defence of the ditch is commonly obtained from small one-tier caponiers, placed where they are least exposed to fire along the ditch, sunk sufficiently to be sheltered from shots descending at 15° from the crest of the glacis, and projecting beyond the escarp only so far as will afford space for two guns or half-a-dozen muskets. The space necessary is more narrowly reckoned in some cases than in others, in order to lessen as much as possible the retirement of the counterscarp opposite the caponiers, which—unless the glacis is raised correspondingly—will make it easier for the enemy to breach the escarp there. A gun can be worked in a width of nine feet, and a machine gun in less; and iron columns can be used, instead of brick walls, to carry the roof; so that the projection of the caponier may be reduced to about twenty feet. But that gives very little room for the detachments, and hardly admits of loopholes for musketry in addition to the gun-ports. A width of thirteen feet (4 m.) for the gunrooms is more convenient, and has been adopted in the caponiers of the French forts. Iron caponiers have been provided for a few German works where it was not possible to screen masonry from curved fire. Apart from their greater resistance when struck, they can be better covered by the glacis, as their total height from the gun-floor need only be about ten feet instead of twenty feet. With wide wet ditches like those of the Antwerp forts, iron, at all events in the form of shields, seems indispensable for the caponiers.

Machine-guns, if they can be relied upon not to get out of order, are more effective for flanking ditches than either guns or musketry; and they take up less space than guns, and require fewer men to serve them. The French have adopted for the new Paris forts a pattern of the Hotchkiss five-barrelled revolving gun, which has a calibre of 1·57 inches, and fires a case-shot containing twenty-four hardened bullets of $1\frac{1}{4}$ oz. each. "The gun is sighted and fixed once for all in the caponier, so that, in a surprise during day or night, it is only necessary to turn the crank, and the gun will discharge sixty to eighty canister shots per minute, consisting of 1500 to 2000 balls."[4] The central caponiers of these forts have three gunrooms on each flank, and as their size exposes them to curved fire, the front walls of the gunrooms are masked by carrying the arches forward about twenty feet beyond them. The casemates in the head of the caponier are also extended laterally, and form orillons sheltering the flanks, so that the plan reminds one of the early Italian bastions.

Brialmont, while adopting "minimum caponiers" in ordinary cases, considers that forts intended to resist a systematic attack to the very last ought to have large caponiers with wide gunrooms protected by masks, and he would give them overlapping heads with acute salients, like those of the Antwerp forts, so that they can be flanked from the ramparts.

The heads of the French and German caponiers are unflanked, and defended only by their own loopholes, which in the case of the former are machicolated. There are counterscarp galleries opposite to them, but these are solely for countermining, and are not loopholed, according to

[4] "Royal United Service Institution Journal," vol. xxiv. p. 287.

Brialmont. As they have no underground communication with the fort, but open into the ditch, it is assumed—surely a very questionable assumption—that men would not stay in them, and that loopholes would be useful only to the enemy. No doubt any reverse defence is likely to fail sooner or later in case of a regular siege, but, as has often been remarked, a fortress has done its chief duty when once it has compelled the enemy to besiege it in form. With detached forts especially, it is assault that it is of most importance to be absolutely secure against; attempts on the caponiers—to blind the loopholes, blow in part of the walls, or smoke out the defenders, would be the accompaniment of any assault, and would largely influence its success; and it seems very desirable to supplement mere direct defence in some way, for the head as well as for the flanks.

"A little while ago," says Colonel Müller,[5] "there was a general predilection for those modes of attack which promised to give escape from a regular siege. But this has lessened with discussion, and consequently the formation of rules for the conduct of the systematic attack has been recently taken up with zeal." Strenuous advocates of the more rapid methods are, however, still to be found. The second part of Major Scheibert's work, published in 1881, "Die Befestigungskunst und die Lehre vom Kampfe," is little else than a vigorous argument in favour of storming the new French frontier fortresses on the outbreak of another war.

In some way or other a place must be got possession of if it blocks the main artery upon which the very life of an invading army now depends, and this, he argues, ought

[5] "Geschichte des Festungskrieges" (1880).

to be done within three weeks from the beginning of hostilities. But the regular siege of a great modern fortress will occupy from three to six months—the duration of a war in these days. The fortress is at its weakest when the enemy first comes before it, and the more promptly and vigorously it is attacked the less are the chances of the defence. The garrison—sure, as Von Scherff says, to consist of second-rate troops, half organized and new to their work—will warrant bold measures against it. If the assailants wait to fortify an investment line, and bring up a siege-train, each day's delay will improve the state of the garrison, and of their works. With an eye especially to Verdun and Toul (of which he gives sketches), he dwells upon the wide intervals between the French forts, and the impossibility of arming and fortifying these intervals, and of clearing away the masses of wood in their front and rear, in the fortnight which there would be for preparation before the German armies appeared. A certain French fortress (Verdun ?) has a garrison of 29,000 infantry. Of these 9000 are required for the forts and enceinte, and 6000 to furnish outposts and guard the intervals on the further side; leaving only 14,000 for the same purpose on the side attacked, or one man to 1·5 mètres. It would be an easy matter, he concludes, for the assailants to break through the intervals at once, especially under cover of night or mist, to establish themselves there firmly, and either push forward directly upon the town, or assault some of the more isolated forts at the gorge while attacking them at the same time in front. The front faces will have about fourteen heavy guns and 250 infantry to oppose such an attack. To keep down their fire, 1250 infantry can be hastily intrenched within 500 mètres, and can be well supported by field artillery. These ought to make it impossible for the gunners of the forts to serve their guns.

Escalade is not so difficult a thing, in his opinion, as people suppose. Caponiers are held in too much awe. It is quite possible to close upon them, and to blind their ports and loopholes; and besides, men when firing through loopholes in masonry are by no means themselves secure from fire, and their defenders, few and isolated, are likely to think more of their own safety than of their duty. But if escalade seems to be impracticable, a lodgment must be made upon the glacis, and mining or heavy guns must be employed to get rid of the obstacles in the ditch.

From the point of view of the attack Major Scheibert's arguments, and his appeals to the experience of recent sieges are very far from convincing; but from the point of view of the defence they are worth bearing in mind, considering the immense difficulties of a regular siege.

THE END.

INDEX.

Abbott, 209.
Aix, Isle of, 125, 139.
Allent, 77, 82.
Angular trace, 117.
Antwerp, 36, 195, 206.
Approaches, 10, 41, 48, 50.
Arago, 156.
Armies, French Republican, 174.
Artillery carriages, 128, 149, 222.
Ath, siege of, 84.
Attack of fortresses, 87.
Augoyat, 13, 65, 78, 111, 132, 148.
Augsburg, 28.

Badajoz, sieges of, 203.
Balloons, 158.
Barras, 179.
Bastioned trace, 28, 37, 57, 116.
Bastions, 26, 38, 89.
Batteries, coast, 124, 182.
———, siege, 43, 52, 85, 150.
Bayonet, 91.
Bélair, 152.
Belfort, defence of, 213.
Bélidor, 129.
Belleisle, Marshal, 129.
Berlin, capture of, 109.
Blinds, 44, 50.
Bois-le-Duc, siege of, 50, 54.
Bonin, 219.
Bruyes, 23.

Breaching, 44, 52.
———, indirect, 194.
Brèse, 205.
Brialmont, 210, 214, 218, 227.
Brisach, siege of, 68.
Brougham, 155.
Brunner, 216, 226, 230.
Buffon, 105.
Bulwarks, 6, 23, 24, 30.
Bureau, 9.
Busca, 44.

Calais, 29.
Cambrai, siege of, 80.
Campi, 41.
Candia, siege of, 77.
Cannon, 6, 11, 53, 105.
Caponiers, 25, 118, 221, 231.
Carnot, 132, 140—144, 155—199.
Casemates, 25, 40, 112, 222.
Cassel, siege of, 201.
Castriotto, 24.
Cavaliers, 27, 122.
Chandeliers, 49.
Charleroi, siege of, 81.
Charles VII., 9.
——— VIII., 11.
Cherbourg, 124, 142, 201.
Chevaux de frise, 55.
Choiseul, 111.
Christine de Pisan, 4, 6, 8.
Citadels, 36, 167.

INDEX.

Coehorn, 81, 83.
Colberg, defence of, 203.
Coni, siege of, 104.
Conti, 102.
Cormontaingne, 129.
Coucy, 6.
Counterguards, 122, 191.
Counterscarp, 21, 190.
Covered way, 21, 190.
Czernichef, 109.

D'Arçon, 143, 147, 202.
D'Aurignac, 78.
Defence of fortresses, 58, 87, 184.
Detached walls, 113, 119, 194.
—— works, 207.
De Ville, 24, 57, 114.
Dime Royale, 95.
Disappearing carriages, 149, 186, 229.
Ditches, 21, 52, 55.
Divisional organization, 178.
Douglas, 194.
Dumouriez, 168, 175.
Durer, Albert, 25.
Dutch School, 48, 54.

Engineers, 15, 35, 64, 92.
——, French, 128, 161, 174.
——, German, 99, 140, 204.
——, Italian, 26, 35.
England, proposed invasion of, 153, 176, 179.
English defence works, 206, 229.

Fausse-braye, 23, 55.
Filley, 111, 132, 147.
Flank defence, 2, 57, 117, 194.
Flanks, hidden, 27, 30, 56.
Flushing, capture of, 183.
Fortresses, application of, 162, 193.
Fort Royal, 121.
Forts, 120, 200—235.
Fourcroy, 111, 126, 132.
Frederick the Great, 140.

French frontier defence, 89, 162, 212.

Gabions, 9.
Gibraltar, defence of, 187.
Glacis, 21.
——, countersloping, 190.
Gneisenau, 203.
Grenier, 128, 136.
Grey, 80.
Guicciardini, 11.
Guise, 17, 30.
Guisnes, defence of, 30.
Gunnery, 53.

Haarlem, siege of, 40.
Henry V., 7.
—— VIII., 28,
Hoche, 176, 178.
Huguenots, 39, 93.

Imaginary attacks, 131.
Institute, French, 153, 180.
Intrenched camps, 90, 204.
Iron armour, 230.

Kars, capture of, 228.
Keeps, 226.
Key of the Treasury, 35.

Lalande, 100, 145.
Landau, 68, 88.
La Noue, 37.
La Vendée, 176, 178.
Lille, siege of, 75.
Lindenau, 140.
Lines of investment, 7, 42, 46.
Louis XIV., 77, 80, 97.
—— XVIII., 197.
Louisburg, 108.
Louvois, 65, 70, 91, 94.
Luxemburg, siege of, 66, 80.

Machiavelli, 14.
Machicolations, 3.
Machine-guns, 232.
Maestricht, siege of, 76.
Maggi, 28.

INDEX. 239

Mallet, 58.
Marchi, 35.
Marollois, 48.
Martini, 21, 26.
Maurice of Nassau, 45.
Mendoza, 43.
Mining, 4, 13, 44.
Mirabeau, 140.
Montalembert, 99—154, 159, 185, 201.
Montluc, 17, 48.
Mortars, 54, 191.

Namur, siege of, 81, 201.
Nantes, edict of, 93.
Napoleon I., 97, 179, 181, 195.
Neuf-Brisach, 89, 138.

Oleron, isle of, 110.
Orillons, 27, 58.
Orleans, siege of, 8.
Ostend, siege of, 50.
Owen, 207.

Paciotto, 36.
Padua, siege of, 16.
Pagan, 57.
Parallels, 77—85.
Paris, fortification of, 90, 204, 210.
Pellisson, 72.
Perpendicular fortification, 117.
Peter of Navarre, 13.
Petersburg intrenchments, 209.
Philipsburg, sieges of, 67, 84, 101.
Pikes, 91, 167.
Pisa, 16, 26.
Polygonal trace, 116.
Polytechnic School, 174.
Prieur, 169, 199.

Ravelins, 27.
Retrenchments, 17.
Revetments, 20.
Ricochet fire, 84.
Robespierre, 169, 176.
Rochelle, siege of, 39.
Rogniat, 204.
Rouen, siege of, 7.

San Gallo, 15, 20.
San Michele, 26.
Sap, 41, 49, 51.
Sappers, 92, 174.
Saxe, 105, 120, 193.
Scheibert, 233.
Shells, 54.
Short service, 166.
Sienna, defence of, 17.
Smoke in casemates, 25, 127.
Soltikof, 108.
Sorties, 86, 186.
Spinola, 46, 50.
Stevin, 56.
St. Simon, 71, 96.
Strassburg, 217, 219.
Surprise, 58.
Swedes in the Seven Years' War, 106.
Systems of fortification, 35, 88.

Tartaglia, 22.
Tower-bastions, 89, 115, 185.
Towers, angular, 119.
Traverses, 223.
Trench cavaliers, 81.
Turrets, 230.

Valenciennes, siege of, 80.
Vauban, 60—97, 133, 138, 158, 185, 102.
Vegetius, 2.
Verdun, 234.
Verona, 206.
Vertical fire, 54, 83, 186, 191.
Vigenére, 22, 35.
Villenoisy, 20, 36.
Viollet-le-Duc, 2, 5, 24.

Wagner, 226.
Walcheren expedition, 183.
Water manœuvres, 56.
Wattignies, 174.
Williams, 42.
Working parties, 45, 51.

Zanchi, 35.
Zastrow, 100.

LONDON:
PRINTED BY GILBERT AND RIVINGTON, LIMITED,
ST. JOHN'S SQUARE.

www.ingramcontent.com/pod-product-compliance
Lightning Source LLC
Chambersburg PA
CBHW031728230426
43669CB00007B/283